A MOVEMENT IN TIME WITH BREITLING AND ROLEX AN UNAUTHORIZED HISTORY

MARK A COOPER

A Movement in time with Rolex and Breitling an unofficial look into the history of the worlds 2 greatest watch producers.

ISBN 0-7414-4168-3

Published by:

INFI(∞)ITY
PUBLISHING.COM

1094 New Dehaven Street, Suite 100
West Conshohocken, PA 19428-2713
Info@buybooksontheweb.com
www.buybooksontheweb.com
Toll-free (877) BUY BOOK
Local Phone (610) 941-9999
Fax (610) 941-9959

Printed in the United States of America

Printed on Recycled Paper

Published July 2007

CONTENTS

This book is dedicated to the most important person in my life, who is my business partner, my best friend, the love of my life and my wife Sandra.

I would like to thank the many dealers and collectors who allowed me to photograph their watches.

Forward

Early watch owners carried a pocket watch for timekeeping. These expensive time pieces became a tool for mostly high achievers (The Rich). Timex was one of the first mass producers of watches, followed by many others. As competition grew among watch manufactures, high quality watches emerged, these where often produced in Gold and became more accurate due to a combination of advances in watch design and production methods.

Some watch companies emerged to become known for accuracy and quality. Many of these companies failed after a few years but the very best still remain with us today, they have lasted over 100 years manufacturing quality watches.

Rolex and Breitling are the World's most recognized watch companies. Known for quality, elegance and the very best in workmanship.

The quality and workmanship comes at a price, so today these watches are still mostly warn by high achievers. As we seldom wear pocket watches, today's wrist watches can do far more than just 'Tell the time". They are now part of most men and woman's dress; many feel naked without a watch.

Since the First World War the wrist watch has taken over. Before the war it was mostly ladies who wore wrist watches, these early models where very unreliable and inaccurate. Most men carried a pocket watch. But the First World War changed people's requirements. Can you imagine the men in the trenches in mud and water having to undo a 'Trench Coat' and layers of clothing to look at ones watch? The majority of watch cases where too large for wrist watches, but as production methods improved it became feasible to produce smaller accurate wrist watches.

Rolex and Breitling became early pioneers of wrist watches. The two world wars played a significant part into the improvements of many of the items we take for granted today such as cars, flight, radio and also watches. Rolex and Breitling produced watches for both sides of the war.

Since the First World war we have witnessed many events in history, Man has been in space, climbed Mount Everest, Ran a mile in 4 minutes, and walked across the poles of the earth. All of these historic events have been achieved by men and women wearing quality watches mentioned in this book. Rolex and Breitling remain at the forefront of these historic events.

Rolex and Breitling produce models and designs for today's fashions but remain the most accurate and highest quality watches throughout the world.

The History of Rolex Watches

Rolex proudly displays the 'Swiss Made' symbol on the watch and now has its head quarters in Geneva Switzerland. Many are lead to believe that Rolex was first made by a Swiss watchmaker many years ago in a little snow covered watch repair shop. Next door, to the toy maker who made the wooden puppet 'Pinocchio'?
Nice romantic story, but that's all it is.

The story of Rolex watches start where they were first made in London, England by British citizen Hans Wilhelm Wilsdorf and his brother in law Alfred James Davis. Wilsdorf was born in Bavaria; he later moved to England in 1900 and became a British Citizen. Wilsdorf was 24 years old. Wilsdorf's money came from inheritance from his late parents, plus money he borrowed from his brother and sister. Alfred James Davis was a self made man. He was already investing in engineering and property throughout London.

The first watches where made at 83 Hatton Gardens, London. Today this is still London's jewelry center and has the reputation as the world's top watch dealer center. The idea of Wilsdorf and Davis was to manufacture wrist watches for men and woman. At the time 1905 most watches where pocket watches, it was mostly ladies who wore wrist watches. After 2 years the company moved into 44 Holburn Viaduct, The property was a two minute walk away from 83 Hatton Gardens, plus it was already owned by Alfred Davis. So he became his own tenant.

Wilsdorf and Davis wanted to make the most accurate watch available, plus include the most luxury. The first watches had the name W & D hallmarked inside the case back. The watches soon become a success, the partners imported many parts from quality case manufactures in Switzerland and other parts of Europe.
They also introduced leather briefcases in a variety of styles.

1905-1908 however was a time when reports of violence and brutalities from Germany started to hit the press. The British who had formally been allies with the Germans, wanted to distance themselves as much as possible. Anything with a German name was changed. The British Royal family themselves changed their surname from Saxe-Coburg to Windsor. The anti German feeling was also responsible for the dog German Sheppard being changed to Alsatian. Doctors stopped informing mothers that their children had German measles, they now called it Rubella.
The watch makers had the same problem; Hans Wilsdorf was a German sounding

name. Alfred Davis made a business decision to change the name of the company. Davis wanted a 5 digit name that could be written across the face of the watch, He liked the name that Timex had. Timex has been manufacturing watches for over 25 years and it was a well known name. The Timex Watch Company made quantity and not quality. It went for the larger market, in doing so its name was well known.

The birth place of Rolex

83 Hatton Gardens, London, England. As it is today. *Circa 2007*

Rolls Royce

A year earlier in 1906 Henry Royce was advertising his new car that he had partnered with Charles Rolls. The Rolls Royce was built as the world's most luxurious motor car. It was said the car was so quite that at 40 mph, you could hear a watch ticking. (40 mph was good for a production car in 1906). The symbol of class, elegance and reliability. This was everything Wilsdorf and Davis wanted to say about their watches.

At an, after performance party at The London Brixton Theatre and Opera House October 1906, Davis spoke with Henry Royce and Charles Rolls, Davis himself owned and drove a Rolls Royce Six Cylinder Silver Ghost. Throughout his life he owned Rolls Royce motor cars. He joked with the car makers, "You may be able to hear a watch ticking in your car, but not one of my watches".

Davis came up with the idea to use part of the Rolls Royce name Known for prestige and quality. Wilsdorf was not happy to change the name.

Most partnerships that have initials as the company name such as DHL, Marks & Spencer and Smith & Western normally put the name and initials in alphabetical order. Not Wilsdorf, it was W & D for his partnership. Even then he had many advertisements as Wilsdorf & Davis. So for Wilsdorf to remove his initial was not a decision he was happy with.

However Davis insisted on the change to continue growth and to disassociate from a German sounding name as much as possible. In the end Wilsdorf gave in, but as you read later, his name once again would appear on many of the company advertisements as managing director. Plus throughout his life he tried various watch companies with his name as the brand name.

Alfred Davis matched the Timex name and Rolls Royce name. The name was more acceptable to Wilsdorf.

ROLLS ROYCE TIM**EX** became the brand we know and love today.

July 2nd 1908 Wilsdorf & Davis register the name **ROLEX**.

Circa 1923

TIMEX and TIME (Horology) where the name came from.

In 197 A.D., Timex, a Roman slave, was supposed to devise a foolproof way of quickly waking his general up or else howling tribesmen would cut his head off on a dawn attack.

Timex came up with a solution to place a candle on the top of a beehive adjacent to his general's bed. When the flame burned down to the hive, the bees became agitated and stung the sleeping general. However, the general was not impressed with the idea, after getting stung several times.

So Timex, whose bee clock didn't work, quickly came up with another invention. He tied lit candles on the heads of goats. The candles were supposed to burn down, jolt the goats from their sleep. However the startled Goats would make such a noise it would awaken those nearby. What would Animal rights advocates have to say about this today?

Timex fact or Fiction would not have been the first recorder of time. Stonehenge in England built 20,000 years ago is believed to measure time, via the sun. The Egyptians built sun clocks as early as 3500BC. In 1656 a Dutch inventor Christian Huygens invented the world's first pendulum clock.

2007 Timex Watch Company produced over 1,500,000 watches per year. They have produced more watches in the world in their 130 year history than any other company.

4

David Plastow Chief Executive of Rolls Royce appears in Rolex Advertisement in 1980. *Circia 1980*

The wrist watch.

Wrist watches became popular during the First World War. In the trenches it was easier to look at a wrist than take off layers of clothing to get at a pocket watch. Leon Breitling had produced a wrist watch for pilots. After Leon's death in 1914 his son Gaston Breitling took over the business, they produced wrist watches for the German pilots, Breitling made the world's first chronograph. WW1 German Ace Pilot 'The Red Barron' Manfred von Richthofen was awarded a diamond crusted Breitling watch.

Rolex was a relatively new company. Movado, Cartier, Louis Brandt (Omega), Timex, Zenith, Minerva, Philippe, Heuer (later to become TAG Heuer in 1985) and Breitling had all been around for a number of years. Louis Brandt (Omega) had been manufacturing watches since 1848. Although it was another German born man Peter Henlein who made the world's first pocket watch back in the 1500's.

Rolex needed to come up with something unique, indeed it did in 1926 when it introduced the world's first waterproof watch case. Again, looking for a name of class, luxury, high society and something that was difficult to open. Wilsdorf and Davis came up with the name Oyster.

At a time of exciting new invention all was not good. In 1933 Ingersoll introduced the 'Mickey Mouse watch. Circa 1932

However today this watch is worth over $2500

Rolex was quick to launch a huge advertising campaign. Plus British woman Mercedes Gleitze, who was the first woman to swim the English Channel, she did so in 1927 wearing a Rolex Oyster. Expensive front page advertisements where taken out in many tabloids around the world. Pictures on billboards appeared with divers wearing the Rolex Oyster. Many stores where even instructed to place a fish tank in windows with the watches displayed working in the water surrounded by Gold fish, and tropical fish. This became so popular that Rolex produced a booklet on how to care for fish, and distributed them to the certified Rolex dealerships.

Circa 1994

ROLEX

Circa 1997

ROLEX

Worth a second glance, even when you know the time.

Oyster Perpetual in stainless steel

Circa 1994

9

GOLD'S FINEST HOURS
THE BENVENUTO CELLINI COLLECTION BY ROLEX

NOBILITY OF DESIGN MERITS A NOBLE METAL. GOLD.
THE CHOICE OF RENAISSANCE MASTER BENVENUTO CELLINI FOR
HIS MOST ACCOMPLISHED ART.
OUR CHOICE FOR A COLLECTION OF TIMEPIECES SCULPTED TO
THE EXACTING STANDARDS OF THE GENIUS THEY HONOR.
CELLINI, BY ROLEX, IN 18 KT WHITE
OR YELLOW GOLD. AS BEFITS
THE LEGACY THEY PRESERVE.

ROLEX
Cellini

WRITE FOR BROCHURE. ROLEX WATCH U.S.A., INC., DEPT. CM, ROLEX BUILDING, 665 FIFTH AVENUE, NEW YORK, N.Y. 10022. WORLD HEADQUARTERS
IN GENEVA. OTHER OFFICES IN CANADA AND MAJOR COUNTRIES AROUND THE WORLD.

Circa 1999

11

The Swiss Connection.

The war (WW1) was expensive for Great Britain, it needed to use all factories for the manufacture of weapons and military vehicles. Plus it made raw materials expensive. So in 1915 the British government imposed a 33.3% duty on all imported watches and parts. To make matters worse, 9 months later it imposed a ban on all imported Gold and Silver.

The war took place mostly in Europe between The allied partners Great Britain, Russia, France and later Italy and the United States, against Germany, Bulgaria, Austria and Hungary. It lasted from July 1914 to November 1918.
A lot of the fighting took place along the border of Switzerland. Switzerland throughout the war remained a neutral country.

Even after the war ended, It was very difficult for Rolex to sell in the lucrative German market whilst the watches where being manufactured in Great Britain. Plus it made matters worse with the high duty and restrictions on Gold and Silver imports to Britain. Wilsdorf & Davis made the decision and relocated to Switzerland in 1919. He located to Geneva, plus formed an alliance with Aegler a Swiss company who became the exclusive supplier of watch movements to Rolex.

Rolex was now free to the build business in a neutral country and sell watches throughout the world. They did however for many years continue to use many British parts such as the Hunter case made by the West Midlands, British company Dennison.

1925 Alfred Davis sold his shares in Rolex. History books simply close the book on Davis after this date. Insiders at Rolex say he had grown tired of Rolex; it was no longer a challenge for him. He and Hans Wilsdorf had different views of the future of Rolex. Davis was a very private man and enjoyed fresh challenges, whilst Hans Wilsdorf wanted his name and pictures everywhere. If he was alive today, he would be compared to other celebrities such as Paris Hilton, (Hilton Hotels) Richard Branson (Virgin Group) Donald Trump (enough said) The Rolex factory had pictures of Wilsdorf everywhere, none of Davis.
So had James Davis simply died after this date? History has no records of what happen to him.
I can't accept that he simply vanished, or moved to the Swiss Alps.
What would a guy do today with all that money from the sale proceeds of his shares, or London Property? I knew he loved the Rolls Royce motor car, so this is where I started my investigations.
In July **1934** a new Rolls Royce Phantom 2 was purchased by a Mr A. J. Davis.
Record at Rolls Royce show it was shipped from England to a tiny village called

Saanenland in the Swiss Alps. Throughout Davis's career with Rolex, he drove and loved the Rolls Royce motor car. This after all is where we have the name Rolex. I have no proof that this is the same man, But I like to believe it was.

Also in the same village a wealthy investor with expert knowledge in engineering funded the world's first drag lift for skiers in 1934. The local hotel owner Oswald Siebenthal and farmer Arnold Annen both put up 33.3% shares in the lift with another unnamed wealthy investor. When the build went 38,000 Swiss Francs over budget. Again the unknown investor stepped in and paid the costs.

38,000 Swiss Francs back in 1934 is close to $24 Million US in today's money 2007.

History cannot take us any further in the life of this great man, but what must be remembered is without Alfred Davis we would not have Rolex today. The watch historians have us believe it was all Hans Wilsdorf. It was not, it was clearly a partnership. Wilsdorf was an excellent manager of marketing, plus he loved the limelight, on advertisements after this date for Rolex watches. They all stated 'managing Director Hans Wilsdorf'. Around the Rolex factory where many pictures of Hans Wilsdorf, some described him as a very vain man. Nevertheless without Wilsdorf we would not have had the Partnership that made Rolex.

May 2nd 1925 Rolex trade marked the coronet (crown). Many writers of Rolex history claim that the famous coronet was not used until 1939. It was infact used much earlier; The Oyster Precision had it on the face. The actual watch worn by British channel swimmer Miss Mercedes Gleitz had the Coronet. This was 1927.

Miss Mercedes Gleitz wearing her Rolex entering the sea before her record breaking channel swim, the swim took 15 hours and 15 minutes. When she came out of the water the watch was perfect and the time was accurate. *Circa 1927*

In **1927** Bell Labourites (The Bell Telephone Company) built the World's first Quartz clock. Head of research and development Canadian born Warren Marrison built the clock based on regular vibrations in the company Labourites in New Jersey USA. Later in 1932 also at Bell labs J.W.A Morrison built the world's first quartz watch and would later prove to be accurate to one second in 30 years.

1940 Rolex started to sell watches in the U.S.A; however they continued to be British Royal Air Force (RAF) pilot's favourite watch. The British government issued a Timex watch to the pilots, the pilots however preferred the Rolex. Rolex or rather Hans & May Wilsdorf introduced a new policy.

<u>NOTICE:</u> ANY BRITISH PRISIONER OF WAR WHO LOOSES HIS ROLEX WATCH TO HIS CAPTORS. WRITE TO OUR GENEVA OFFICES AND WE WILL REPLACE THE WATCH.

The above was Hans Wilsdorf's idea. Wilsdorf and his wife May, personally handled the correspondence, until her sudden death in the winter of 1944. The only clause was that after the war the British soldier would repay Rolex. The payment also had to be made in Swiss Frances, although after the war any currency could be used.

Circa 1943

Rolex and the British Prisoners of war.

Rolex and the "Prisoner of War" watches is a fascinating story, but a story that has become almost a legend in retail sales methods. If you were a British prisoner of war captured by the Germans during World War II, you could write to Rolex, Geneva and they would send a watch to you free of charge. The reality is only a little different; it seems that this service was only available for British prisoners. The letter would be sent to Rolex from the camp via the International Red Cross, who (like Rolex) was headquartered in Geneva. Hans Wilsdorf himself or his wife May, who wrote a letter that accompanied every watch dispatched to a P.O.W.

The first letter is to a senior officer in the RAF, who is now a guest of the German Air Force. This was because one of the stranger German habits during WWII was that each service ran its own prison camps "catering" to prisoners from the opposing service. This all can be gained from the address "Stalag Luft 3", Stalag, meaning "prisoner of war camp" and Luft being short for "Luftwaffe" or Air Force.

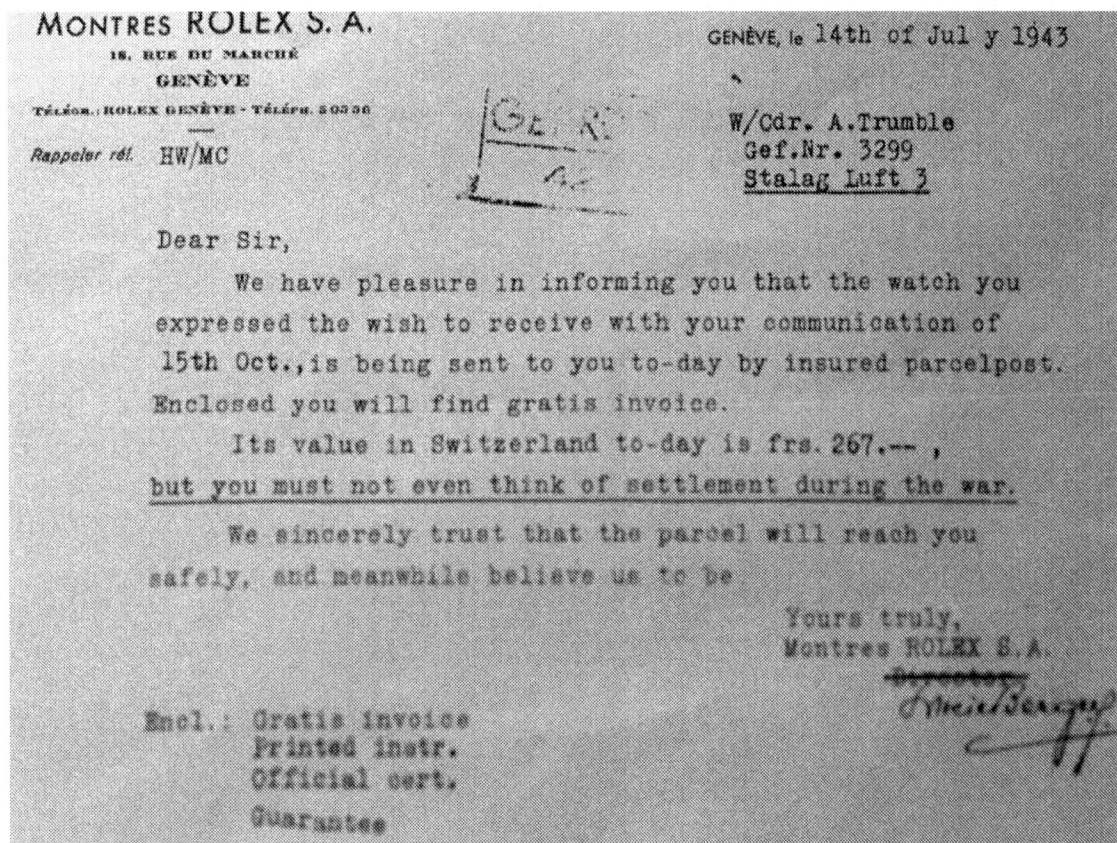

Circa 1943

Montres ROLEX S.A.

15, RUE DU MARCHÉ
GENÈVE
TÉLÉGR.: ROLEX GENÈVE - TÉLÉPH. 500.56

Rappeler réf. HW/MC

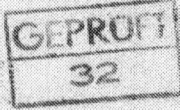

GENÈVE, le 30th of March 1945

GEPRÜFT
32

Ger.Nr.
Stalag Luft 3

Dear Sir,

We beg to acknowledge receipt of your order dated
10th March 43, and in accordance with your instructions we
will supply you with 1 Chronograph Oyster No 122.
This watch costs to-day in Switzerland Frs. 250,—
but you must not even think of settlement during the war.

As we have now a large number of orders in hand for
officers, there will be some unavoidable delay in the
execution of your order, but we will do the best we can for
you.

Meanwhile, believe us to be

Yours truly,
Montres ROLEX S.A.
Director
H. Wilsdorf

Circa 1945

MONTRES ROLEX S.A. GENÈVE

ROLEX WATCH C₀ LTD.

Genève, le 7th of Sept. 1940

Rappeler Réf. HW/MC

Dear Sir,

We have received news from Germany that Grahame is in
good health and quite happy. We sent you a telegram already yesterday
reading as follows:

"Grahame well address No 1085 Oflag VII C/H Germany"
if prefer sending us your letters will readdress— Rolex Watch Co"

so that you should be informed at once. We are looking after his wants
in the same way as for some other British Officers, who are also
prisoners in the same camp. Please rest assured that we will do everything in our power to obtain food or other articles which he might
be particularly keen to receive.

If you want to communicate with Grahame through our firm,
please do not hesitate to send any letters to us and we will reforward
immediately upon receipt.

Meanwhile, we beg to remain

Yours truly,
Rolex Watch Co. Ltd.
Director

*All letters to us should be
sent by air to save
several weeks delay*

Circa 1940

16

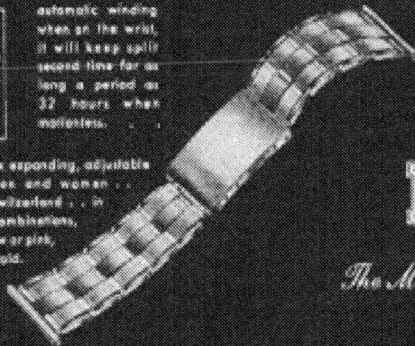

Circa 1975

17

After May Wilsdorf death, Hans Wilsdorf created the *Hans Wilsdorf Foundation*.

It was a trust that specified how the Rolex Watch Co should be run after his death. Thus ensuring the company would never be sold and never publicly traded. Profits from the company to this day are allocated to charities.

From the early 1940's North and South America became a new and very lucrative market for Rolex. The Datejust line was introduced and became a favorite for the U.S market. Rolex even named one watch the 'Thunderbird' this was named after the U.S Air Force aerobatics team. That particular watch only enjoyed limited success. The Thunderbird pilots themselves preferred Breitling watches.
In **1945** Rolex introduced the 'Tudor' watch in an attempt to catch some of the lower priced watch market. Tudor initially had an English Tudor rose as an emblem, but in 1960 changed to shield. It was then like it is today known as a cheap Rolex, although today many are sold as Rolex/Tudor.

Rolex still kept its very strong British connections; Prince Charles was given a watch by Rolex for his 12 birthday. In **1953** Rolex sponsored the British expedition to conquer Mount Everest; the summit was reached 29th May. However to Rolex embarrassment it was later discovered that Sir Edmond Hilary wore a Smiths watch. The Smiths watch was made in Streatham, South London, England.

1954 Rolex introduced the highly popular Oyster Perpetual Submariner. This was waterproof to the depth of 600ft. (182m). At the same time it produced a deep sea watch that could still work at 10,335 (3,150 meters). The Cyclops was first introduced. The bubble on the crystal was first fitted this year to the Datejust.

1956 Rolex changed the Datejust to display day and date.

Andre Heiniger took over the company when Hans Wilsdorf passed away **July 6th 1960** at the age of 79.

The **1960's** was a great time for Rolex. The company grew and now had World wide acclaim. The Help of free advertising helped. The famous French diver and Television personality, Jacques Cousteau wore a Rolex whilst filming under water scenes. The British films James Bond, all featured Bond wearing a Rolex. The astronauts in space wore Rolex. However without gravity the perpetual self winding system failed and they had to manually wind the watches. U.S President Franklin D Roosevelt wore a Rolex.

There's only one time World Champion
Jackie Stewart takes off his Rolex.

When he puts on his fireproof underwear.

This only happens when the world's premier auto racer dons his famous tartan helmet and climbs into the cockpit of his Tyrrell-Ford.

No matter, though. There's scant time to take his eyes off the track when blistering down the straightaways at speeds exceeding 160 mph.

The race over, Jackie Stewart's Rolex is right back on his wrist.

Our craftsmen are honored that a man whose life often hangs in split-second timing picks Rolex for his personal timepiece.

Such confidence is not misplaced.

Everything about a Rolex brings it to the pinnacle of the watchmaker's art.

The Oyster case is hewn from a solid block of gold or stainless steel.

The patented Twinlock winding crown screws down onto the case (similar to a submarine hatch) to provide an utterly secure barrier against water and dirt.

The Rolex crystal is meticulously diamond-cut for a mirror-perfect match with its case. And ingeniously designed so that it actually seals tighter under pressure. (Deep underwater, for instance.)

Into this practically impregnable case goes the peerless Perpetual movement.

Each is made the Old World way. With pride. And patience. By dedicated Swiss watchmakers, heirs to a tradition of excellence.

One at a time, the movements are hand-tuned to exceptional accuracy.

A gravity-powered rotor (invented by Rolex) means that wrist movement does the winding; the wearer never needs to give it a thought.

Every single Rolex movement is submitted to one of the impartial Swiss Institutes for Chronometer Tests for 15 full days of rigid trials. Only on passing does it win the coveted "Chronometer" rating. (Although Rolex accounts

for only a tiny fraction of all Swiss watch production, nearly half of all the chronometer certificates ever awarded have gone to Rolex.)

Even so, Rolex' own inspectors then put each one through final tests before giving it their stamp of approval.

A lot of work? Yes. But that's what makes a Rolex a Rolex.

That's why Jackie Stewart prizes his. And why sportsmen and adventurers are invariably Rolex men. And why Rolex is the official timepiece of Pan American World Airways. And why most of the world's heads of state, men who have scaled the heights, proudly pick Rolex over all others.

Rolex. A sign of a special kind of man.

You'll feel it the minute you wear a Rolex of your own.

Each Rolex earns the recognition it enjoys. You know the feeling.

ROLEX

The Rolex Oyster Day-Date is 30-jewel, self-winding chronometer (with date available in 18 languages) in 18 kt. gold, $1080 with bracelet at life. Also in 18 kt. white gold or platinum.

American Rolex Watch Corporation, 580 Fifth Ave., N. Y., N. Y. 10036. Write for free color brochure. Also available in Canada.

Circa 1973

19

Circa 1937

The **1970's** brought the start of probably the worst events for Rolex, Rolex dealers, buyers and collectors. THE FAKES. I have devoted a large chapter to this area that to this day is one of the largest problems we have. Also throughout the 1970's and 1980's came the new watches from Japan. Seiko and Citizen produced inexpensive quality quartz watches. Their design broadly based on Rolex and Breitling watches. They hit the watch stores and jewelers across the world, they produced the watches in very high numbers, and achieved in a few years that previously took watch makers decades to achieve.

Rolex sponsored the world famous Wimbledon tennis tournament, held in South London, England in **1978** and has done so every year to this day.

By the end of the millennium Rolex had a value of £2. Billion pound Sterling or $3.6 Billion U.S.

Rolex watch profiles

1910 Rolex awarded the first official chronometer certification. This was the first award given for a wrist watch.

1926 Rolex invents the World's first waterproof case. The 'Oyster'. The Oyster is made using 904L Steel for the case. Rolex is the only watchmaker that insists on using this for its steel cases. 904L is an extremely tough super alloy. A huge 250-ton press is required just to stamp out a single Oyster case from a solid block of 904L. It seems no measure is too extreme for Rolex.

1928 Rolex introduced the 'Prince' The most accurate watch in the World at its time. This was just 1 year before the 1929's stock-market crash. Innovative in design and engineering at the time, it was made ``for men of distinction,'' according to the advertisement. The vintage Prince disappeared in about 1950. Rolex rolled out a modern version in 2005 for its 100th anniversary. These early Rolexes aren't waterproof and should be wound daily. However they remain a classic and collectors watch, at auction they fetch $12000-$35,000 today.

1931 Rolex Perpetual was introduced. The Worlds first automatic winding mechanism
1933 Rolex introduces the 'Bubbleback' watch. The watch was also the first to have the words Rolex Oyster on the dial. All dials had 'Swiss made' at the 6 hour.

Circa 1933

Circa 1933

Early Rolex Bubbleback.

1945 The Rolex 'Oyster Perpetual Datejust'. The Worlds first watch to display the date. Without any doubt today the most recognizable watch. Various models over the years have been produced. The Jubilee Datejust was one of the first. It was only available in 18kt gold. Since 1982 the Rolex Datejust model has been fitted with a sapphire crystal. These are produced in a laboratory and have the Cyclops fitted over the date. The two-tone Datejust is the top selling Rolex watch. Since its launch it has been the favorite and the style is now copied by Citizen, Timex, Seiko and Ingersoll.

1953 May 29th. The Rolex Explorer was introduced. Given to the British team that conquered Mt Everest. (However this was NOT worn by Sir Edmond Hilary) This is easily recognizable with its black dial and large luminous triangle marker at 12 and luminous Arabic numerals for the other quarters. These watches can withstand extreme temperatures of between -20 c and + 40c without changes of in its viscosity. The Explorers hands and hour markers are made in White Gold, and painted with White Tritium.

Everest Leader's Tribute to Rolex

Circa 1954

The first Explorers were produced before the conquest of Everest. The name
Explorer was Registered in January of this year. Rolex did later produce a dress
Explorer. This was really a standard Oyster Perpetual in steel with a white dial. Not
many where produced? Later in the 1960 they produced the Date Explorer, again
only in steel as a dress watch.

1953 Rolex introduced the 'Rolex Turn-O-Graph' this features a black rotating bezel to measure elapsed time.

Turn O Graph *Circa 1962*

1953 Rolex also introduced the very popular 'Oyster Perpetual submariner' the divers watch.

Submariner *Circa 1953*

1954 The Milgauss was introduced by Rolex. It was designed for people who work in power plants and research laboratories, where electromagnetic fields cause the watch to dysfunction. The movements that where mostly alloy where encased in an iron shield. The demand was very low; because of this they are very rare and are prized by collectors.

Milgauss *Circa 1954*

1954 The 'Cyclops' was first fitted to the 'Datejust'. Still used today to clearly display the date, this has become a symbol of the Rolex watch. Also in the year the Submariner was first produced. It was an overnight success.

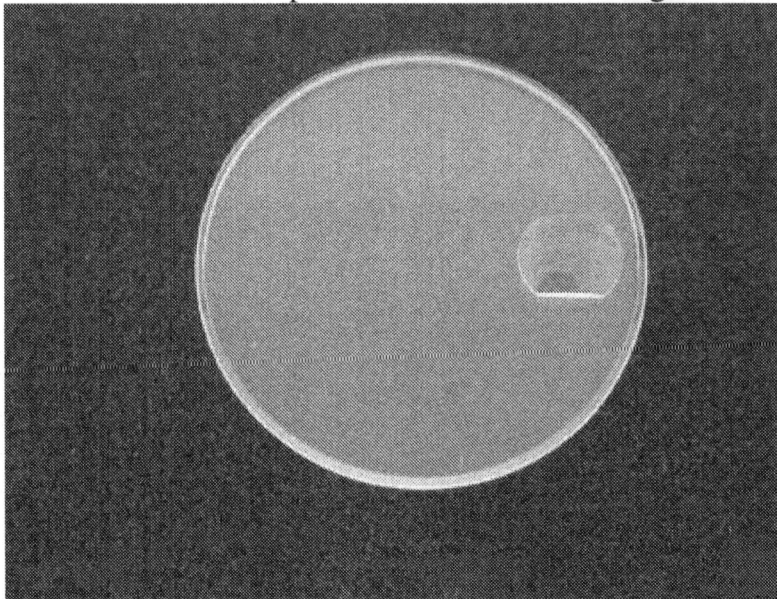

Cyclops *Circa 1954*

With the Rolex Sea Dweller that came out in 1971. Rolex took a large part of the divers market. In 1965 it became the British Royal Navy's official military issue watch. Whilst the British Royal Marines ordered Submariners. It is also known as the 'James Bond 'watch. The watch could dive to a depth of 660 feet. Today the Submariner and Sea-Dweller are guaranteed to dive down to 2000ft, (600m)

1955 Rolex introduced the G.M.T-Master. This watch was the first to display the exact time in two different time zones. This was to coincide with pilots and passengers flying across the globe. Pan Am requested the watch from Rolex. Many of the Pan Am pilots were upset by the move as they preferred Breitling. However Rolex and Breitling signed a contract to become the official watch of Pan Am. In 1973 General William Sewell the CEO of Pan Am appeared in one of the Rolex advertisements. The G.M.T-Master has a fourth hand and is able to keep time in 2 time zones. The red color fourth hand is used with the rotatable 24 hour bezel. NASA astronaut Jack Swigerts wore a Rolex G.M.T Master on his flight on the ApolloXIII. This actual watch is now mounted on a plaque on the wall of Rolex Geneva HQ. Along with signed pictures of the other astronauts, all Rolex owners. G.M.T is named after Greenwich Mean Time, the world's standard time. The G.M.T was powered by a new movement, the caliber 1565. Today over 50 later the G.M.T is still a top selling Rolex watch. It is purchased by aviators and people who require knowledge of multiple time zones. Many currency traders on Wall Street wear the watch for this reason.

The GMT-Master *Circa 1955*

"When you love what you're doing as much as I do, it doesn't feel like work." *Plácido Domingo*

Wherever he travels, Plácido Domingo carries a series of green bound books into which he writes his engagements up to three years in advance. Such are the demands the opera world makes on one of its most sought-after performers.

His ability to thrill an audience is such that a legendary curtain call in Vienna lasted over an hour. "It would have been easier," Plácido said, "to sing the opera all over again."

Domingo has committed nearly a hundred operatic roles to memory. He believes this daunting repertoire is necessary to attract the widest possible audience because his ambition is to help more people, all over the world, enjoy the music he loves.

More than a singer, Domingo is also a respected conductor. "The operatic conductor is like a Roman charioteer," he says. "He has a hundred horses on stage and a hundred horses in the pit. And he has to control them all."

Throughout his career, Domingo has held himself to the highest standards of performance. It's the same measure he uses when selecting a timepiece. "My Rolex is perfect for me," he states. "You could say it's one of my favorite instruments."

ROLEX

Circa 1997

1956 The 'Oyster Day-Date' was introduced. Also known as the 'President' the watch as the title says, displays the day and the date. It was available in 26 different languages. Worn by American President Dwight D Eisenhower. When the watch was first produced many in Geneva question whether or not it could keep accurate time, due to amount of mainspring energy required to move both the day and date disc simultaneously. However Rolex proved the doubters wrong. The watch was a major success. The watch was also worn and owned by President Franklin D Roosevelt, Kennedy, Johnson, Nixon, Ford and Clinton. President G W Bush wears a Rolex Prince.

Circa 1990

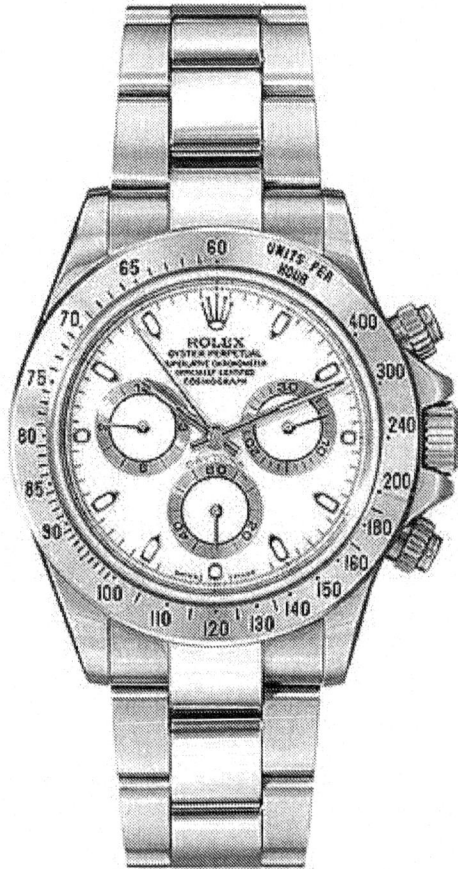

Circa 1960

1960. The Oyster 'Daytona' Cosmograph. This watch was a wrist watch and a stop watch. Also known as the 'Paul Newman' after he wore this watch in the film 'Winning'. It was named in Honor of the racing drivers who competed on the world famous Daytona race track in Florida. The drivers appreciated its instantly readable dial.

1967. The divers watch known as the 'Sea Dweller'. Capable of diving 2000ft (600m) was first launched. The watch was an instant success and is still a top selling model 40 years later.

Circa 1970
The Rolex Sea- Dweller

1970 The 'Rolex Oyster Quartz' made its début. Many in the trade did not like the idea that Rolex known for mechanical watches and accuracy would produce a quartz watch. But Quartz are so accurate, they had to bring out a line of quartz watches. The worst thing for Rolex was the amount of fake watches that came on the market. The first production of the quartz models where sold out immediately. However overall the quartz was not a good seller for Rolex. The first Quartz watches had the sapphire crystal and were the first Rolex to have the hacking feature. The watch also had the quick change date. Unlike Breitling, Rolex never made any watches with LED displays. If you see one it's a fake.

Despite what you may think, back in 1966 it was infact Rolex that was involved in the development of the quartz watch. Via a consortium of watch manufactures that included almost every watch company, but Breitling. The consortium consisted of Rolex, Rado, Enicar, Longines, I.W.C, and Bulova. Omega, Zenith, Movado, Le Coultre, Zodiak and Patek. It was founded January 1961 and produced the first working prototype in 1966.

The prototype was known as Beta 1. In 1967 Beta 2 was entered in the Geneva chronometer tests. It has a precision score of 0.152. Considering the best ever recorded mechanical performance was recorded at 1.73. A perfect score would be 0. The Quartz was a breakthrough, over 10 times more accurate than mechanical movements.

It was such a difference that the consortium put pressure on the Council at Geneva to allow it to be entered. After 1968 quartz watches have been banned from entering the competition. It was changed to mechanical movements only and still is today.

Rolex Oysterquartz Datejust *Circa 1988*

1971 The 'Oyster Perpetual Explorer 2' this watch had a fourth hand with Orange arrow to indicate night and day.

Rolex Oyster Perpetual Air-King *Circa 1976*

32

Circa 1976

Rolex Perpetual Air-King Movement

1988 The 'Oyster Perpetual Daytona' was reintroduced. This time with a perpetual rotor. The watch is produced in a variety of material, 18Kt yellow Gold, 18kt White Gold and a stunning Steel and White gold model. Steel dials, Black dials, Black mother of pearl with roman numerals, a fantastic looking Meterite dial (Light two-tone blue) or Diamond paved dial with blue sapphire hour markers. Sapphire Bezel, with blue calf hide leather strap. Below

Circa 1999

1991 Rolex finally released the new version of the Daytona. They had abandoned the workhorse Valjoux movement in favor of the Zenith movement. However as from 12th June 2000 Rolex started to produce their own movement for the Daytona.

1992 The Rolex 'Yacht Master' was introduced. Available in 18ct yellow gold only. In 2001 they then produced them in platinum as well.

The Oyster Perpetual Yacht-Master is a handsome watch for business and social wear, yet remains, first and foremost, a Yachting watch.

The automatic movement with perpetual rotor has obtained the official title of Swiss Chronometer. It is housed in the famous Oyster case.

The Triplock crown, a triple safety device, is protected against accidents by seamless shoulders cut in solid metal. The date changes at midnight and can be rapidly adjusted. The bracelet has extension links so that it can be worn over a light diving suit and is secured by a Fliplock safety clasp. A tough sapphire crystal remains scratch resistant under the most arduous conditions. The new Crystal is also fitted to the Sea Dweller. It is guaranteed waterproof to a depth of 300 m (1,000 feet). The revolving bezel incorporates a ratchet and can only be turned anti-clockwise. It is equipped with a Fliplock bracelet. Oyster Perpetual Yacht-Master models in 18 ct gold and combined metals are available with either a black or a blue dial with matching bezel.

Circa 1992
Rolex Platinum Yacht-Master

2005. The Rolex '**Prince**' is reintroduced. It had been nearly 80 years since the original 'Prince' was introduced. The crystal back showing the movement makes it popular. The Rolex Prince attacks the strong hold that Cartier had on quality dress watches.

Prince William was given a Rolex Prince as a Birthday gift. William exchanged it for a Rolex Sub-Mariner. The official response was that The Sub-mariner being waterproof and sports style was more fitting for an active young man. However Buckingham insiders say that he did not like the name Prince on his watch. William changed it to a Sub-Mariner just before his basic training at Sandhurst Military Academy in the UK.

Official press release picture of Prince William wearing his Rolex Sub-mariner
Circa 2005

The Rolex Prince Cellini is now the top selling Rolex dress watch. It is now available in 18ct, Yellow, White or Everose Gold. The Prince is still hand wound like the original Prince, but is now waterproof. The new Cellini range includes the Prince, the Quartz powered Rolex Orchard, the hand round Rolex Cestello, the Quartz powered Rolex Danaos mage in Pink Gold, the Quartz powered white gold Rolex Cellissima, and the men's hand wound Rolex Cellini Cellinium.

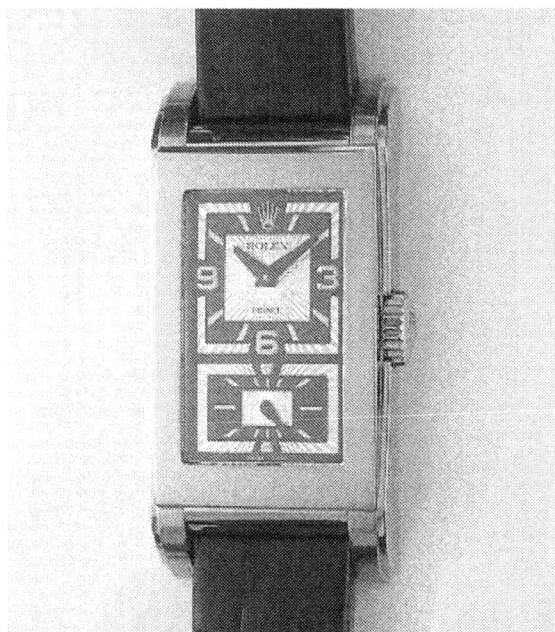

Circa 2005

The new Rolex Cellini

Part ll

Fakes, Copies and Replicas.

However the manufacturers wish to call them, they are all fake. Some now advertise as **Genuine Swiss made Rolex Replica**. Whichever way you call it, they are all fakes. A genuine Rolex replica is just a fake whether it is manufactured in Switzerland, China or in Grandpa's tool shed.
Since the 70's Fake Rolex watches have flooded the market. Today there are so many they are actually graded 1-5. 1 being a good copy, to 5 being very poor cheap imitation. The dealers don't care who they scam.

In China its big business to produce Rolex watch boxes, certificates of authenticity and warranty certificates. They even produce and sell the plastic hang tags. They

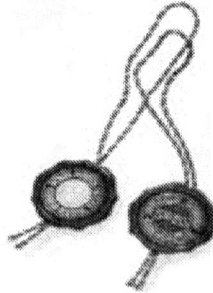

sell them on sites such as eBay.   Fake Rolex hang tags.

Whilst mentioning the eBay company. The same buyers that buy these fake boxes, tags and certificates. Go on and sell watches as Genuine Rolex watches. Just do some research into the buyer and see what he has been buying. Buyers who then go and buy a watch from a guy who a few months earlier bought fake boxes, deserve all they get. Sorry to be blunt, but please ALWAYS do your research when buying a Rolex, Breitling, Ball, Tiffany, Movado or Cartier watches. The above watches are the most copied watches, However today even Seiko and Citizen, fake watches are produced.

The next few page contain tables of the 5 typical grades of fake Rolex watches, the table can be also used for fake Breitling watches.

Grade 1	Replica, Fake watches by grade.
Make	Made in Switzerland, So the genuine Swiss made markings are correct!
Movement	ETA up to 25 jewels
Hacking Signal	Yes
Gold Finish	2 Tone, Solid Gold, All Gold, Triple wrap, Gold plate.
Color of Gold	The colors of grade 1 Replicas are almost exact, some are the exact color.
Stainless Steel	Solid and good quality grades used.
Weight	Almost exact, Only a very fine set of scales can differentiate.
Water Resistance	Normally to 100ft (30m)
Crystal	Genuine Sapphires used. Some are very good quality.
Wholesale price	$495-$800
Retail Price	$685-$1500
EBay Price	$1200-$12,000 As they are sold as genuine Rolex.
Warranty	1 year if purchase from a replica Rolex website. From eBay Zero warranty.
Comments	These are really nice watches; they look good and feel good. It's easy to see why there is a market for them.

Grade 2	Replica, Fake watches by grade.
Make	Made in Japan.
Movement	ETA up to 25 jewels
Hacking Signal	Yes
Gold Finish	2 Tone, Solid Gold, All Gold, Triple wrap, Gold plate. Singe wrap.
Color of Gold	The Dress models are a very good color match, but the Submariner is a bad color match.
Stainless Steel	Solid and low quality grades used.
Weight	Typically 15-25% lighter than the genuine Rolex.
Water Resistance	Ok to take a shower, but not in the pool. Poor quality seal.
Crystal	Genuine Sapphires used. Low quality.
Wholesale price	$295-$600
Retail Price	$485-$795
EBay Price	$1200-$12,000 As they are sold as genuine Rolex.
Warranty	9 months -1 year if purchase from a replica Rolex website. From eBay Zero warranty.
Comments	These are nice Japan made watches. They are as good as a Casio.

Grade 3	Replica, Fake watches by grade.
Make	Made in Japan
Movement	Copy movements, normally made in China. They will state 25 Jewels, but this is just wording.
Hacking Signal	Yes
Gold Finish	Gold plate.
Color of Gold	The gold is a very brown color.
Stainless Steel	Solid low grades used.
Weight	15-25% Lighter
Water Resistance	Splash proof
Crystal	Mineral
Wholesale price	$149-$249
Retail Price	$299-$495
EBay Price	$1200-$12,000 As they are sold as genuine Rolex.
Warranty	6 months if purchase from a replica Rolex website. From eBay Zero warranty.
Comments	Probably as good as a supermarket watch or what you find in a shopping mall. Expect it to last 1-2 years max.

Grade 4	Replica, Fake watches by grade.
Make	Made in Japan, Korea, and China.
Movement	China made movement, some quartz.
Hacking Signal	Some are.
Gold Finish	Thin Gold plate.
Color of Gold	Too shinny
Stainless Steel	Plated
Weight	Can be heavy due to Steel case, Chrome plated material
Water Resistance	Ok if you are caught in the rain.
Crystal	Mineral
Wholesale price	$25-$70
Retail Price	$149-$249
EBay Price	$1200-$12,000 As they are sold as genuine Rolex.
Warranty	3-6 months
Comments	Poor quality watch. Easy to establish this is not a genuine watch, It will also have poor markings.

Grade 5	Replica, Fake watches by grade.
Make	Made in China / Hong Kong
Movement	Made in China
Hacking Signal	No
Gold Finish	Fake Gold plate.
Color of Gold	Too Shinny
Stainless Steel	Plated.
Weight	As light as a feather
Water Resistance	Don't break into a sweat.
Crystal	Plastic
Wholesale price	$10-$29
Retail Price	$75-$150
EBay Price	$895-$5,000 As they are sold as genuine Rolex. If you follow the steps later in the book, this won't be you.
Warranty	1 day if you can find the seller
Comments	They are often sold to drunk Germans and Brits on vacation in the south of Spain by Moroccans.

Rolex Fakes and Replicas, How to spot them.

This next section is dedicated to spotting a fake Rolex. Each section has a point of reference, learn each point and when you examine your future watch, go through every point.

Let's get the easy bit out of the way. Apart from the Oysterquartz and Rolex Tru-Beat the Rolex second hand does not tick around the face of the watch. This was always an easy way to spot a fake. Plus the misspelled words. But now the producers have got wise and produce some very convincing watches.

The worst place to get taken in by the scammers is on Amazon.com, eBay.com, yahoo.com and Craigslist.org.

eBay

If we stick to the World's largest internet auction site, eBay, I will explain how you can avoid buying a fake. Although to be fair to eBay, they are doing everything they can to stop this. They do not allow sellers to sell any fake or replica items on eBay. If you see one, on the item page you can click on a link and report it. eBay will look at the item and if it is fake they will remove the listing and send a warning to the seller. If the seller continues, they will be permanently removed from eBay.

Yes, you really want that great looking Rolex on your compquter screen, and the price is perfect. "Wow, what a deal"
Now go to the bathroom take a break, and think about it.

1) How long has the seller been listed with eBay. You had better hope more than 12 months. If not look elsewhere.

2) Does he or she have good feedback as a SELLER? Not someone who has been buying $0.99 items such as e-books. (Books or pictures that can be downloaded.) Or other low cost items, just to boost feedback ratings. Do some research, you are possibly about to buy a good watch and save some money. We all work hard for our money so it's worth it for you to spend time looking into the seller's feedback history.

Recently I saw a fake Breitling get sold on eBay for $3600. The seller had a feedback of 6 and had previously purchased a used pair of jeans, and used child's car seat. Ask yourself? Would this same person also own a $5000 Breitling watch! Plus the seller would not take Pay Pal (eBay's on line payment system). The seller wanted a money order, basically cash. I could tell by the pictures it was fake, but not everyone can. By the time you have read this book you may be able to spot a fake yourself.

Only buy from a buyer who has been SELLING for over 12 months and has a feedback rating of 99% or more.

3) ONLY pay by Pay Pal. If the watch turns out to be a fake and you can prove it with documentation, you can return the watch and Pay Pal will refund you your money. You may just have to pay to ship the watch back to the seller. But a complete refund you WILL get. However this must be claim within 30 days of payment for guarantee of full return, you may get some back after 30 days, but this is not always the case.

4) Inform the buyer before you bid that you intend to take the watch at your expense, to a Rolex dealer to get the watched cleaned and serviced. If it turns out to be a fake. Inform him you will return the watch for a full refund.

If it's genuine the seller has nothing to worry about, and the cost of having a service on your new Rolex is well worth it. Plus you have complete peace of mind.

5) Never, EVER buy from a seller who is based in China, Hong Kong or Africa. It's so frustrating to see sellers in China selling genuine Rolex and Breitling watch straps. Come on think about it, these are made in Switzerland, why would they be less expensive in China?

6) If it is offered always pay for the insurance.

7) On eBay there are many sellers who have written guides on Rolex, plus there is a discussion board site. There is a section for most things including a site just concerning watches and clocks. If you are concerned about a particular watch for sale copy the item number down, and ask the opinion of others who use the site about the item. I often answer question on the site as well as other watch experts. Here is one occasion where the word Replica is used in a Rolex Advertisement. It refers to a ladies watch, being a replica of the man's. *Circa 1948*

8) The crooks will sometimes end an auction early; this can be done at anytime by a seller who wants to accept the amount bid so far. This stops eBay from shutting the auction down, once a fake has been spotted and reported.

9) Once you get your watch examine it carefully and look at the pictures on the site. I have seen genuine pictures copied and pasted from genuine sellers, for example a genuine Rolex Datejust may have sold last week. The thief's will sell a fake Rolex Datejust and use the pictures from the genuine watch. This is hard to get over, however if you follow all the rules and **only** pay by Pay Pal you will be OK.

Circa 2000

1980's Rolex Oyster DayDate *Circa 1981*

Above, a complete fake Rolex with fake tag and fake green sticker. It's a pretty poor fake, but if you bought it on the internet, its looks real enough. Please beware.
At the time of writing there are over 100 websites selling 'Genuine Replica Rolex's'. (fake) along with replica Sunglasses and handbags. They set up in various locations across the world and are hard to track down. However if you buy a Replica item you risk being sued for thousands of dollars and /or thrown in jail.

The Second Hand Store or Pawn broker.

You are more likely to get a genuine watch here as the stores owners have seen hundreds of people come in there store an offer a fake watch. Most staff and owners know how to spot a fake. I buy watches from many pawn brokers, but always get a return form authorized. You can pick up a real bargain here. Many are genuine Rolex watches. They end up at the Pawn broker for all sorts of reason, death, divorce, loss of employment or just financially troubled.

On the next page is the form I produced and use. I would NEVER buy a watch from such a store without getting this signed. Ensure you are dealing with the owner or person in charge before you complete the form and purchase your watch.

You can use something like this.

I ………………………………. Agree to purchase ………………………………… …

Serial number……………………………for $………………………………….......

From………………………………...

....

I will within 14 days have the watch serviced by a certified Rolex dealer at my own expense. The seller above agrees unconditionally to fully refund my money without any charge if the watch is not a genuine Rolex as stated by the seller.

Signed……………………………….Buyer….dated…………………………….

Signed……………………………….Seller….dated…………………………….

Feel the Quality of a Genuine Rolex.

Rolex produce some of the best watches in the world. Every detail is perfect. I would recommend you visit an Authorized Rolex Dealer. Feel, hold and touch the watches.
Look and feel the strap and clasp. Put one on, do it up and undo it. You will learn how they feel and get an idea of the weight. Yes the sales person will try to sell one to you. Go alone and tell them you will get your wife/husband/mom or dad to help you choose. That way you will not feel awkward when you want to leave without making a purchase. (Sorry if you are a watch sales person, but we have to try)

A Rolex is very heavy compared to a $20-$100 watch. The Oyster cases are made from a solid lump or ingot of stainless steel, gold or even platinum. However some of the better fakes are Genuine Rolex Replica as they illegally call them do use solid gold cases.

Look at the Color of the watch if it is gold. The early now vintage Rolex did use 9kt and 14kt gold. The new Rolex uses 18kt. The fakes are plated so they have a brassy color (Too Brown). Look at the color of something that is 24kt gold. It's a darker brassy brown color. The plated fakes look like this.
The bezel is normally a giveaway, but you won't know what I mean until you have done some research and seen your fair share of fakes versus the genuine watches.

The Conversions are very difficult to spot. A conversion is typically example is where a genuine stainless steel watch is turn into a two-tone model. This can easily be by replacing any of the genuine items with fake bracelet center links with cheaper gold plated or solid gold links. I have even seen fake bezel, crowns and hands fitted. Some of the solid gold bracelets and bezel are normally 9ct Gold, if you scratch it, it's still solid gold. I have also seen lower quality diamond dials and bezels. You are still buying a genuine Rolex and real diamonds (Sometimes) and solid Gold, but filled with non genuine parts to enhance the watches value.
Another point to know is that the date wheels fitted to Stainless steel models are either White or Silver. The Gold and two-tone (Gold and Stainless steel) are fitted with a Gold color wheel. This is an indication of a fake or a conversion.

Below is a Replica/Fake Yacht-Master. I took this picture directly over the watch. Look at the Date 6th. On a genuine Rolex, This will be exactly central in the Cyclops. This is an easy give away. Also this fake has the word Rolex and Cornet printed on the face/dial. It should be slightly raised.

Circa 1990

FAKE/REPLICA

An instant giveaway is the Cyclops; it's not central and not magnified to 2.5 times.

ROLEX Yacht Master

Circa 1990

Above is a genuine Rolex. Look at the date 1st. It's as it should be central in the Cyclops.

Rare Rolex GMT with clear crystal back case!

I spotted this for sale at a jewelers, I had never seen such a watch. The jeweler informed me it was very rare and worth every penny of the $5500 he was asking for it. I thought I knew a lot about Rolex watches and was annoyed that they had produced a model that I had never come across before. I was ready to write out a check, what a find this was. He explained he bought it from a young man whose grandfather had died and left it to him. However I like to look a little deeper before I part with my money. I spent hours on research looking into this. Rolex has never ever produced such a watch. I went back and asked to look at the watch again. The model number was 16660. *See Case numbers quick reference guide.*

That number was for the 1978 Rolex Sea Dweller Submariner. I told him without any doubt it was fake. I offered the jeweler $35 for it, to use as research for this book.
He refused and was pretty upset. He had paid the guy $1200 for it. He said the story about the sellers Grandfather sounded true and he thought he would make a nice profit. The markings on the watch where good apart from this.
Maybe it was the guy's grandfather's watch, and poor grandfather died thinking he was leaving his grandson a nice watch. It could have ended up being passed down for generations to come. None of them ever knowing it was a fake. However more likely the guy told the jeweler a nice story. Either way, you can see why the producers of fake and counterfeit goods must be stopped. The Jeweler still has the watch today, as a personal reminder to look into things a little deeper before parting with his money.

The Rolex Daytona are beautiful watches, they are the second highest fake watch produced after the ever popular Rolex Datejust. If you look at a Daytona it has a luminous square on every hour hand except for the 12 position. There are currently 2 producers from Japan who make a very good Fake, Both have luminous marks at the 12 o'clock position. A huge mistake by them and easily spotted if you know.
On Submariners and Datejusts, they do not have a luminous marking where the date is.
Back cases. The Rolex watch in most cases does not have any engraving or logos on the back, apart from Rolex Sea dweller produced after 2002 had "Rolex Oyster original gas escape valve".
Some rare ladies 1940's watches did have "Original Oyster case by Rolex Geneva and a tiny coronet" stamped on the back case. Apart from these do not expect to see any markings on a genuine watch.
The fakes back cases get quite inventive, they often will have a Large Coronet

(Crown), plus model numbers, serial numbers and /or hallmarks engraved on the back cases. I have even seen some molded back cases. However Rolex do NOT do this, it's a sure sign of a fake.

The green sticker on the case back:

If you buy a new Rolex, I would advise you remove it and stick it to the certificate. If not it will in time wear and fan anyway.

The Green back case stickers with a Hologram, Rolex coronet and model number are applied to new Rolex watches.

Of course the counterfeits do not have these, the latest fakes I have seen just coming out in the middle of 2007 have the holograms too.

If you are buying a used Rolex, The sticker was probably removed, so if it does not have a sticker, it does not mean it is a fake.

The Cyclops

Most genuine Rolex watches produced with date apart from the Rolex Sea Dweller have the Cyclops; this is the glass bubble that sits on the Crystal (Glass front) directly over the date. This was first fitted to Rolex Datejust watches in 1954. It has since become a signature for the Rolex line.

It magnifies the date, thus making it easier to be read. It magnifies the date and makes it 2.5 times bigger. (2.5 x Magnification). You can look at the bubble at any angle and still be able to read the date.

Lower grade Cyclops has 1-1.5 magnification. The difference may seem small, but when you are familiar with a genuine Rolex Cyclops the difference is obvious. Plus remember the Rolex is a precision made watch. The bubble will be directly over the date, and the numbers below will be dead center. On lower and medium grade fakes it will be off center. Some of the highest grade replicas now do have 2.5 times magnification.

If buying via the internet, ensure you have a picture from directly over the watch so this can be checked.

The Crown (winder)

The Rolex Submariner, Daytona and Sea-Dweller are produced with the unique Rolex 'Triple Seal Crown'. It has an extra seal to prevent water seepage. It has an extra seal within the Crowns threaded tube. I have only ever seen 1 very good Grade 1 replica with this. Plus some replicas made with some genuine Rolex parts. But you are only likely to see genuine Rolex parts on the Very high priced Rolex and Breitling watches such as The Oyster Day-Date Platinum or Gold with bezel set with 53 diamonds. Of course the replicas do not have real Diamonds. Plus the beautiful Lady Datejust in 18kt gold, set with 40 Baguette rubies and 10 diamonds. Another good check is to unscrew the crown, it should be tight to unscrew, but run at an even pressure (smooth). If it is loose and makes a scratch or tinny sound, it is possibly a fake. Genuine Rolex and Breitling watches are smooth and quiet.
Also try and set the time (Rolex) turn the hands back and forth. Look closely at the hands, if the second hand bounces or jumps around erratically it's probably a replica.
While we are looking at the crown (winder) look at the Coronet. Rolex make them in one solid piece of metal. A replica/fake sometimes has the coronet glued or soldered on.

Additional Features

Closely look at the extra features such as those on the Daytona. The Daytona (also known as the 'Paul Newman Watch' after the American Actor wears this watch in old films and everyday life). Has small sub dials on the watch face. The replicas have these but very often they do not function.
The genuine Rolex second hand will go right up to the second and minute markings. Plus on a genuine Rolex the minutes hand is rounded, not just cut flat. Many fakes have incorrect hands, which are often the wrong shape and size, but unless you have the genuine watch with you are you really know your Rolex watches you can get caught.
Get a Rolex brochure if possible with the watch you want.

The Hacking Feature

In 1972 Rolex introduced a new feature, called the Hacking feature. Pull out the crown to the second click (to change or set the time). On a genuine Rolex produced after 1972 the second hand will stop. See the table on previous pages on the grades to distinguish which grade of fake has this. The lower grades simply continue ticking.

Date Dials/wheels

The date numbers seen under the Cyclops should ALWAYS be perfectly centered in the opening, with NO exception. They should always be perfectly clear. I always do a quick test and using the winder stroll through the numbers to ensure every number is clear and perfectly centered. Remember Rolex do not make seconds. They make the highest quality watches that you will find anywhere in the world.

Overall appearance and Markings.

The replicas are good at this, but they are not perfectionist. If you look at a Genuine Rolex closely against a Replica, The genuine watch markings are perfect. The Fakes/replicas have pitted markings. The Coronet on a genuine Rolex Dial is clear and has no scratches, or marks on the painted letters. The hands on a Rolex that are painted are perfect and painted to the edge. You will see many fakes that have an outline on them. As you are most likely to just have the one watch in front of you, it is very important to cover every detail, point by point, and so you can verify you are looking at a genuine Rolex.
Swiss made; on the watch dial/face either side of the 6th hour. A genuine Rolex produced after 1999 has the words Swiss made. Previous to 1999 the genuine Rolex had either SWISS-T< 25 means less than 25 milliCure of Tritium   or T SWISS MADE T or T SWISS T for radioactive material Tritium, with a maximum of 25 milliCure.   Pre 1950 radium was used, although these older watches just had 'Swiss Made'
There are more details later in the book under Glossary.
Another overall appearance is the inside edge of the watch. This applies only to Rolex watches. Look at the watch at a slight angle, the inside edge of a genuine Rolex has a satin finish. Most of the grade 1-5 replicas I have seen all have a polished mirror like edge. You can see the hour markings reflected in the fakes. The polished edge looks nice, but Rolex do not have this. This can be an early warning sign of a fake.

Genuine Rolex, Note the case number between the lugs. **5500**. If you follow the case number chart in a later chapter, you will note it is for a Rolex Explorer Air King.

Box Pillow Cushions

A genuine Rolex box will have a pillow cushion which your watch is wrapped around. This keeps protect the bracelet. The pillow are typically an imitation leather/suede, they are soft to touch, like velvet. The fake/Counterfeits are cheap not only in appearance, but also to the touch. The genuine Rolex pillow is produced in Sweden and has the Rolex coronet logo embossed on the front.

Rotating Bezel

Another check is the Bezel if it has one. The Rolex sports models have them. The Rolex Submariner, Sea Dweller and of course the GMT-Master all have a rotating bezel. The genuine Bezel on the Breitling and Rolex watches will click around the watch 120 times. It has very fine ratchet movement; it will be smooth and click around 2 clicks to every second. The fakes click at 60 clicks, and will be loud and harsh.

Serial numbers

The Rolex Oyster watches case will have a serial number engraved between the strap lugs at the 6 o'clock position. At the 12 o'clock position it will have a case reference number.

Certificate/Paperwork

As discussed earlier fake certificates can be purchased quite openly on eBay. Today we have good quality color photo copiers and this helps them with their counterfeiting. Rolex and Breitling watches produced after 1990 have certificates, but these certificates have a water mark, a Coronet for Rolex and Breitling wings logo on the Breitling paper work. The certificate should have the watch serial number. Rolex also have a red seal plus a code only identified by those who know Rolex.

Now you will know. The red letters printed and embossed on the warranty card will identify when the watch was first sent to the authorized Rolex dealer.

For example code 'W-CR' means the watch was sent June 1991.   6- 91

R	O	L	E	X	W	A	T	C	H	E	S
1	2	3	4	5	6	7	8	9	0	11	12

Take a copy of this code; it's a very powerful tool against fakes. It's not widely known. If you go to buy a Rolex from a private seller, dealer, second hand store, pawn broker and they say genuine with paperwork. Check the date. I have recently, it was a 1995 watch, but the number on the certificate was O-HL that would mean Feb 2003. Before you even look at the watch you know it's a fake.

Rolex fights back against the fakes.

Rolex has spent nearly 100 years making watches and making itself a worldwide brand name. It is fighting back against the counterfeits. The company works with many government trade standards as well as Interpol.

Plus it is getting clever at staying 1 step ahead of the fraudsters.

Evidence is the new Hologram green stickers on the back of new watches, with individual serial number. Plus since March 2002 the crystal (glass) has micro-etching. A tiny Rolex coronet is etched onto the glass at the 6 o'clock position. It's so tiny; many of you who have a genuine Rolex would not have noticed it. As it's hard to see with the naked eye. Once you have seen one and know how faint and small this is, you will look out for it. Again before you buy, you really need to go to a Rolex dealer to examine a genuine watch first.

I hope this section on fakes helps. Read all sections and use each section as a point. You need to look at every point before you buy.

May 10th 2007 An American business man Mike Korpi from Portland Oregon went on a trip to Shen Zhen, China. While there he bought 8 fake Rolex watches, he paid $14.40 each for them. A total $115.20. The watches where for himself and his children and grandchildren. US customs found the watches, 1 he was wearing and 7 he had in his baggage. He was fined $55,300 for bringing in counterfeit watches into the US. This would be the value of genuine Rolex watches. Korpi aged 55 will be 67 before the fine is paid off, The US government will garnish his wages.

Rolex can also take proceedings against him, and they can ask for damages of $100,000 per watch, however this is thought unlikely.

May 15th 2001 online auctioneer EBay revealed that two of its European subsidiaries have been sued by watchmaker Montres Rolex S.A. and some Rolex affiliates for alleged trademark infringement. Rolex alleges that EBay subsidiaries EBay GMBH and EBay International AG have been, "infringing Rolex's trademarks as (a) result of users selling counterfeit Rolex watches," through EBay's German Web site, according to EBay's quarterly filing with the Securities and Exchange Commission (SEC). The lawsuit, which was filed in the regional court of Cologne, Germany, also alleges unfair competition, EBay said. The company added that Rolex is seeking an order forbidding the sales of Rolex watches on the EBay Web site as well as damages.

This never happened, but it shows how far Rolex will go to protect its trademark. Now eBay will not allow ANY counterfeit goods to be sold, whether it is watches, shoes or DVD's. They work with manufactures and honor trademarks. Today eBay is a safer place to buy a Rolex and many Authorized Dealers do sell on eBay.

Case numbers quick reference guide.

Rolex watches have a case number, its often known as the model number and sometimes called the serial number. Whatever name you want to give it, this number will help you identify your watch. But most importantly help fight against fraud/fakes. This book is an unofficial guide; Information is very limited from Rolex. There is no public record.

In producing this guide Rolex have only offered very limited resources. Please use it simply as a guide. More information may be added to another edition.

Circa 1932

Case Number	Watch/Style
60	Chronograph
871	Chronograph Moon Phase
971 A-U	Prince
1002	Air King
1004	Zephyr
1016	Explorer
1019	Milgauss
1020	True-Beat
1072	Speed-King
1074	Chronograph
1223	Chronograph
1343	Prince
1400	Air-King
1401	Air-King
1406	Submariner
1427	Explorer
1490	Prince Century
1491	Prince
1527	Prince Railway
1530	OP Date
1541	Prince
1550	OP Date
1564	Prince Sporting
1573	Imperial

1599	Prince Sporting
1600, 1601, 1603, 1604, 1605, 1607, 1610, 1611, 1620, 1622, 1623, 1624,	Datejust
1625	Datejust T-Bird
1626,1630	Datejust
1650,1651,1652	Cosmograph
1655	Explorer II
1656	Cosmograph
1657	Explorer II
1658, 1659	Cosmograph
1660	Sea Dweller
1661	Submariner Date
1662	Yacht-Master
1665, 1666	Sea-Dweller
1670, 1675	GMT Master
1671, 1676	GMT Master II
1680	Submariner Date
1768	Prince
1770	Prince JH
1802-1839 *and any between*	Day-Date
1852, 1873, 2240	Bubbleback
1862	Prince Railway
1871	Prince Sporting
1873	Bubbleback
1894, 1895	Master Day Date
1936	Egyptian

2021, 2022, 2023, 2057, 2226	Chronograph
2081	Scientific
2240	Bubbleback
2280	Royal Observatory
2303	Chronograph
2319	MW Royal
2490	Bubbleback
2507, 2508	Chronograph
2518	Egyptian
2574	Chronometer
2595	Royal Speed King
2730, 2771	Prince
2737, 2811	Chronograph
2764, 2764	Bubbleback
2917, 2918, 2920	Chronograph
2940, 2949	Bubbleback
2942	Scientific
3009, 3019, 3065	Bubbleback
3036, 3055	Chronograph
3078	Chronometer
3082, 3084, 3085	Chronograph
3116, 3117, 3120, 3121	Speed King
3130-3135	Bubbleback
3138-3139	Army
3330, 3333, 3335,3346	Chronograph
3347, 3348, 3353, 3358	Bubbleback

3359	Viceroy Sky Rocket
3361	Prince
3372	Bubbleback
3386	Royal
3458	Bubbleback
3462, 3481, 3484	Chronograph
3474, 3478	Centregraph
3525, 3529	Chronograph
3548, 3549, 3595, 3598, 3599	Bubbleback
3635, 3642	Chronograph
3646	Divers
3668	Chronograph
3693	Speed-King
3695	Chronograph
3696	Bubbleback
3716	Athlete
3725	Bubbleback
3765	Royal
3767	Bubbleback
3794	Lifesaver
3795, 3796, 3801	Bubbleback
3827, 3834, 3835	Chronograph
3868, 3882	Precision
3877	Empire
3938	Prince
3940, 3951	Bubbleback

3978	Pioneer
3997, 4048, 4062	Chronograph
4070	Falcon
4099, 4100, 4113	Chronograph
4125	King of Wings
4127	Athlete
4191, 4219, 4222	Precision
4220	Speed-King Royal
4270	Elegante
4311, 4313, 4332, 4352	Chronograph
4361	Speed King
4365	Precision
4376	Prince Railway
4392	Bubbleback
4402	Prince
4444	Air-Giant Royal
4467	Datejust Big Bubbleback
4500, 4537	Chronograph
4547	Date
4556	Precision
4647	Observatory
4767, 4768, 5036	Chronograph
5045, 5048, 5050, 5051, 5052, 5055	Bubbleback
5056	Speed-King Kew
5068	Prince
5087, 5105, 5173, 5488	Bubbleback

5100	Quartz Date
5500, 5501, 5504	Explorer, Air-King
5502	Air King
5505	Everest
5506	Explorer
5508, 5510, 5512, 5513, 5514	Submariner
5520	Air King
5700	Air King Date
6006, 6007, 6011, 6015, 6016	Bubbleback
6020	Precision
6021	Speed-King
6022	Air Giant
6031	Big Bubbleback
6032, 6034, 6036	Chronograph
6048, 6052	Bubbleback
6062	Moon phase
6065, 6065, 6074, 6075. 6076	Bubbleback
6087	Bubbleback
6098	Explorer
6104, 6105	Datejust
6106	Bubbleback
6144	Royal
6150	Explorer
6155	Datejust
6200	Submariner
6202	Turnograph

66204, 6205	Submariner
6210	Kew Observatory
6232, 6234, 6236, 6238	Chronograph
6239, 6240, 6241	Cosmograph
6244	Royal
6251	Datejust
6262, 6263, 6264, 6265	Cosmograph
6266	Date
6270	Chronograph
6299	Explorer
6304, 6305	Datejust
6309	Thunderbird
6342	Speed-King
6350	Explorer
6418, 6421, 6430	Speed King
6420, 6422, 6423, 6427	Precision
6466, 6480	Precision
6510, 6511	Day Date
6512	Veriflat
6516, 6517	Datejust
6536, 6538, 6538A	Submariner
6541	Milgauss
6542	GMT Master
6543, 6544, 6545, 6546	Oyster Perpetual
6547	Oyster Date
6548-6554	Oyster Perpetual

6556	Tru-Beat
6558-6580	Oyster Perpetual
6581	Oyster Perpetual Octagonal
6582	Oyster Perpetual Zephyr
6582-6599	Oyster Perpetual
6602, 6604, 6605, 6609	Datejust
6610	Explorer
6611	Oyster Perpetual Day Date
6612, 6613	Day Date
6614-6623	Oyster Perpetual
6624	Datejust
6625	Oyster Perpetual Date
6627	Datejust
6634	Oyster Perpetual
6664	Date
6694	Date Precision
6706-6771	Oyster Perpetual
6800-6839	Datejust
6862	Yacht-Master
6900-6928	Datejust
6929-6935	Yacht-Master
6944	Oyster Perpetual Date
6962	Yacht- Master
6994	Date
7131, 7169	Chronograph
7205	Divers Watch

7575	Oyster Perpetual
7576-7600A	Oyster Perpetual Date
7603-7811	Oyster Perpetual
7815-7928	Datejust
7972	Oyster Perpetual Date
8011, 8012	Air-Giant
8015	Drake
8023	Canadian
8027	Viceroy
8029-8045	Datejust
8050, 8052	Explorer
8053, 8055, 8058	Oyster Perpetual Canadian
8056	Bubbleback
8060	Oyster Perpetual
8065, 8066, 8067	Datejust
8074, 8075, 8076, 8077,8079, 8080	Oyster Perpetual Canadian
8171	Moon Phase
8180, 8237	Chronograph
9162	Chronograph
9210, 9211, 9220, 9221	Oyster Perpetual
9659	Date
14000, 14010	Air-King
14060	Submariner
14203-14238	Oyster Perpetual
14270	Explorer
15000-15505	Oyster Perpetual Date

16000-1624	Datejust
16520, 16523, 16528	Cosmograph
165590, 16570	Explorer II
16600, 16650, 16660	Sea Dweller
16610, 16613, 16618	Submariner Date
16700- 16760	GMT Master
16800-16808	Submariner Date
17000, 17013, 17014	Oyster Quartz Datejust
18000-18389	Day Date
19018-19168	Oyster Quartz Day Date
67180-67518	Oyster Perpetual
68158-69159	Datejust
69160A	Oyster Perpetual Date
69163-69188	Datejust
69190A	Oyster Perpetual Date
69198E, 69198R, 69198S	Datejust
69240	Oyster Perpetual Date
69258	Datejust
69268	Datejust
69279	Datejust
69279A	Datejust
69288	Datejust
70114	Oyster Perpetual Date
70886-70991	Oyster Perpetual
81014, 81015, 81015R, 81022	Datejust
96887, 96887A, 96888, 96891	Oyster Perpetual Date

Rolex Watch Database--Stolen ROLEX

The Rolex database includes dates of service when a watch has been worked on at one of two authorized Rolex service centers – New York and Dallas.

Also included in the database is the serial number of watches reported stolen.

If a stolen watch is brought into an authorized service center for service, it will be confiscated. Information from the database is available to persons with legitimate need to know, including law enforcement, insurance investigators, jewelers and others.
Telephone requests are not honored.

Myth buster **There** is no master registry of Rolex ownership. A website should be checked out www.watchsearcher.com
This is the Swiss Official Database of stolen watches. This covers Rolex and Breitling.

United States
Rolex service center 9420 Wilshire Blvd, Beverly Hills CA 90712
Rolex Watch Service Corp, Rolex Building 2651 North Harwood Dallas TX 75201 214-871-0500
Rolex Watch USA, Inc. Rolex Building 665 Fifth Ave New York NY 10077
Rolex (Fax) 212-980-2166
Rolex: Telephone 212-758-7700
www.rolex.com

United Kingdom
THE ROLEX WATCH COMPANY Ltd.
19 St. James's Square
London SW1Y 4JE
Tel. + 44 (0) 207 024 7300
Fax: + 44 (0) 207 024 7317
www.rolex.co.uk

Australia
Rolex Watch Australia Ltd. 70 Colin's St Melbourne Victoria 3000
Tel. + 61 3 9654-39-88 Fax 61 3 9650-44-99

Tudor Watches and other Rolex models

In 1934 -1935 Rolex was number 1 in high class watch manufacture in the world. It was now producing the Oyster and had a worldwide market. Hans Wilsdorf wanted Rolex to grow bigger. With Davis out of the picture Wilsdorf was free to pursuit more ventures. He experimented with 'The Wilsdorf Watch'. He contemplated registering his name. He did earlier in 1932 register the name 'Wilsdorf Watch Company"
However anything with his name would have to be the best. Since Rolex was already the best quality watch on the market such a watch would only attack his own market. He changed his mind and went for a lower quality watch, which would not affect the Rolex brand.
He used a variety of Watch Company names, Rolco the name came from **Rol**ex **Co**mpany. Marconi Watches, Elira Swiss Watches, the Brex Watch Co and Unicorn Watches.
All these watches where to be second class to Rolex, they were produced with less Jewels and movements from companies such as Rebberg. At this time Rolex was using Aegler movements. It was Hermann Aegler who purchased Alfred Davis Shares. Herman Aegler was appointed chairman of the Rolex Watch Company.
The main problem Rolex had with the subsidiary watches was how to promote them without damaging the prestige brand name Rolex.
The subsidiary watch brands where produced in very large quantities until 1945.
Tudor (The surname of the English Royal family up to Queen Elizabeth I) It did have the English Tudor rose as its first symbol. However in 1963 this was changed to a shield.
Many consider a Tudor to be a 'Cheap Rolex'. It is true that they have different movements that Rolex use. However they are also high quality. Tudors are manufactured under the same strict guidelines as Rolex and the guarantee is the same.
I have yet to see a Tudor fake; I don't think they have ever been produced. A Rolex Oyster can fetch far more than a Tudor Oyster. Rolex are one of the world's most accurate watches, but it still has it status, and this is why many are sold. A Tudor is just a nice accurate watch. If you are going to spend this sort of money a Breitling or used Rolex could be purchased. However on eBay you will see many sellers selling Tudor watches as Rolex/Tudor. They fetch a higher price this way. Early Tudor watches are now a collectable item as they are rare. Many Tudors made with the same case as Rolex typically have fewer jewels.

Circa 1990

Tudor Prince Oysterdate.

So who buys Rolex watches? Tudors?

In a 2007 British Survey for the Daily Mail newspapers financial supplement, The Financial Mail. The following results were recorded.

48% of BMW drivers, own a Rolex.

28% of Mercedes Benz drivers, own a Rolex.

If this is correct 76% of Rolex owners drive a BMW or Mercedes. What car do you drive?

After this you have the following well known celebrities, who are Rolex owners.

Paul Newman, Elton John, Roger Moore, Prince William, Prince Harry, Robbie Williams, Tom Jones, Tiger Woods, Brad Pitt, Eddie Murphy, AC/DC guitarist Angus Young, Neil Diamond, Ex President Bill Clinton, The late JFK and many more.

Breitling celebrity owners include John Travolta, Mel Gibson, Richard Branson, Puff Daddy and Arnold Schwarzenegger.

19 % of commercial airline pilots own a Breitling.
15% of commercial airline pilots own a Rolex.
9% of commercial airline pilots own an Omega.
8% of commercial airline pilots own a Citizen watch.
2% of commercial airline pilots own a Cartier watch.

A Rolex President, given to President JFK by Marilyn Monroe and inscribed with the worlds "Jack with love always from Marilyn May 29th 1962". This was a birthday gift to him when she sang her world famous 'Happy Birthday' song. In October 18th 2005 the watch fetched at auction just over $120,000.

A good price for a $12,000 watch, However April 2006 a Rare 1941 Rolex Medical Chronograph 'Anti-Magnetic' fetched $1,385,000 US, at auction in Geneva.

Where not to wear a Rolex! Maybe it's just me as I am a watch collector, but I always look at a person's watch. If someone is trying to sell me something it sends out 2 messages. 1, This person is successful and should know what he is selling or 2, This man/woman is making a fortune selling this, and is just after commission!

Depending on what they are selling puts me on guard and probably many others. When I was a very successful salesman, I never wore a Rolex or drove a large German car. I prefer to tone it down and turn up in an average new car such as a Ford and wear a good watch such as Citizen or Seiko. However if you are selling million dollar homes, a Rolex and German car will show off your success.

I have friends who work with their hands, but are Rolex enthusiasts. Some of these wear a fake Rolex at work and wear the real thing after work. Personally I hate the fakes and would never wear one.

Some watch collectors will wear a Tudor watch for this purpose. It was advertised in the Horological Journal back in the 1950's as the watch you could 'punish' without mercy. If your aspirations are higher than your bank balance. It is still marketed at anyone who leads an active life, such as Police officers, and blue collar workers. It is also given as a graduation gift, as a 'starter Rolex'.

Current Rolex Models

The current Rolex collection include Oyster Perpetual Air King, Date, Datejust, Day-Date, Oysterquartz, Submariner, Explorer, Explorer II, Sea-Dweller 4000, GMT Master II, Yacht-Master, Cosmograph Daytona, Rolex Prince, Lady Oyster Perpetual, Lady-Date, Lady Datejust and Lady Datejust Pearlmaster.

Before leaving Geneva, every Rolex watch must travel through a high-tech obstacle course of quality-control checks. Every dial, bezel and winder will be checked and double-checked for scratches, dust and aesthetic imperfection. The microscopic distance between its hour and minute hands will be painstakingly calibrated to ascertain that they are lying perfectly parallel. An ominous-looking air-pressure chamber will verify that each watch is waterproof to a depth of 330 feet. (The Submariner and Sea-Dweller divers' models are guaranteed to 1,000 and 4,000 feet, respectively.) And every watch will engage in a precision face-off against an atomic-generated "überclock" that loses but two seconds every 100 years. Only after successfully passing dozens of checkpoints does a watch receive the Rolex seal.

Co-Branded Rolex

Over the years Rolex has Co-Branded many watches for company awards, although these are rare, they do not fetch high prices on resale markets. Below Coca Cola, Next page Dominoes Pizza and Winn-Dixie.

Circa 1993

Circa 1993

Collecting Rolex Watches.

Due to the costs not many are able to buy a genuine Rolex watch. While writing this book many non Rolex owners have expressed to me that Rolex is a watch for the rich. I could not disagree more with that comment. I believe it's a watch for a high achiever, those with ambition and have worked very hard in life to get where they are today. It is the people with no ambition who do not own a Rolex.

Collecting Rolex must start somewhere. For many it's a Tudor, used or fake Rolex. It need not be. A used Stainless steel Rolex air king can be purchased for as little as $1,000 US. The new price for these watches is in the $2,000 US area. It's a great all round watch and perfect for day to day use. However you can pick up early Rolex Explorer watches for around the same price. The Explorer first came on the market in 1953. So if you are looking at a 50 year old watch. I have seen many 1960 models on eBay selling for approx $1200US. These can be cleaned and have new crystal's and bracelets fitted if required. Instantly you have a classic vintage Rolex watch. However the most popular watch is the Datejust.

An early Datejust with the Cyclops can be found in stainless steel and gold bezel for approx $2200.

Once you have 1 Rolex many later decide to either upgrade to a higher quality Rolex or get another. The Rolex prince comes high up on the order list when purchasing a second watch. The watch is the ultimate dress watch. Also the Rolex Daytona is normally purchased by existing Rolex owners.

Some of the Diamond encrusted Bezel on white Gold or Platinum Dayton's cost up to $250,000 new. Although they do start at $20,000.

I have seen many Rolex collections and would like to thank the owners for allowing me to photograph many of the watches. No one collection is the same. Some collectors prefer pre 1960's and W & D watches, others prefer to collect pocket watches and some will collect watches of all brands and colors of dials. Some high achievers will collect every brand produced by Rolex.

When you see a private collection like this, it's like walking in a jewelry store; obviously security is at its highest. Some collections are kept in well lit wine cellars'.

The average collector (if there is one) has 3 or 4 Rolex's. Often another classic will have found its way into the collection such as a Breitling Bentley or Omega dress watch. Always in the collection an original Rolex Oyster can be found.

Choosing where to buy your Rolex is important. Yes you can get some great deals at Garage sales, (Car Boot sales to UK readers), Flea markets, second hand stores and pawn brokers, Public auctions, Jumble sales, Charity stores such as Goodwill or Oxfam. However do you know what you are looking for, and can you now spot a fake? Go over the how to spot a fake section and start looking in quality stores getting to know the feel of your Rolex.

Today Rolex is still manufactured in the Bienne area of Switzerland, is headquartered in Geneva, and is sold all over the planet. Wherever you may be in the world, Rolex watches are associated with impeccable quality, prestige, and luxury. Hans Wilsdorf and Alfred Davis, would, no doubt, be very proud.

Pink Gold and Steel 1945 Rolex Perpetual Bubble Back. *Circa 1945*

1953 Rolex Stainless steel Explorer I
Circa 1953

Solid 18 ct Gold Rolex Daytona Pearl Dial 2005 model *Circa 2004*

Circa 1995

Now if you read the first section, the history of Rolex, you will remember that it was not started by a watchmaker in a little watch makers store in Switzerland.

The opposite is true of Breitling.

What are similar are the ages of the men. Hans Wilsdorf was 24, so was the man who started Breitling.

It was in the tiny village of St Imiers, Switzerland that 24 year old Leon Breitling started the Breitling watch company in 1884. The village was in the Jura Mountains, at certain times of the year the only access was on skis and snowshoes. Leon Breitling was a certified watch maker at the age of 19. He started his career at age 6. Yes, 6. His parents where farmers. However like most people living in St Imiers in the 1870's they worked at home building clock parts. The local factories where small, and to increase the size and heat would have been very expensive, plus due to the weather and poor roads, many workers would have failed to make it in to work every day.

The Breitling family had watch parts delivered, where they cut, painted and assembled small parts. Leon was fascinated in the art of clock manufacturing. Whenever his father would go to the local factories, Leon always went along. Whenever new items that the Breitling family had not seen before to be assembled or machined, Leon would often play sick so he could stay off school and work on the clock parts. At 15 he went to work in a watch factory.

Leon Breitling's first tiny factory started producing Chronographs and Precision counters for scientific and industrial purposes. Leon used home workers to help produce parts, just as he and his family had done. Many of these Leon had known over the years, or he had either met them at Church, or at the local clock companies Christmas party. The workers respected Leon; he was after all one of them. Leon valued his work force and at the beginning tried to visit many of them. This way he could demonstrate the precision and quality of workmanship he wanted.

Leon invested all his savings into his small company, many weeks he went without any income from selling his clocks and watches. He had to personally ask his cottage workforce to continue unpaid until the sales came through. He worked from 4am to 11pm for over 3 years building his company. At times he had to personally deliver items to his cottage workers; on occasion on foot many miles through snow covered lanes.

His workforce respected Leon Breitling and produced components for him unpaid until he had sold his products. It only happened twice at the start of his company that he could not afford to meet labor costs. However eventually all his workers where paid plus they all received a bonus for their help.

Included in the Chronographs made were instruments for cars and later Gliders.

1891 saw German Otto Lilienthal fly more than 165 ft in his glider. His glider had a tiny wind counter, made by Leon Breitling. Lilienthal was known as the father of aviation. This was forgotten when some 11 years later the American Wright brothers flew 'The flyer' due to the hatred of the Germans in the early 1900's, Otto Lilienthals flight seems to have been forgotten. Like the Wrights. Otto Lilienthal worked with his brother Gustav Lilienthal. But ask yourself, when you think about the birth of aviation you would never think of the Lilienthal Brothers?

Otto Lilienthal in 1891 12 years before the Wright brothers famous flight *circa 1891*

Circa 1919

12 months after Lilienthal's famous or not so famous flight, Leon Breitling moved his factory to La Chaux-de-Fonds, in those days this was the center of Swiss watch making. The Breitling watch company grew and built a reputation of precision watch makers.

1909 July 25th Louis Bleriot flew his airplane Bleriot XI across the English Channel from Calais, France to Dover, England in 37 minutes. He was known as "Conqueror of the Channel" His watch was a Breitling, and he carried a Breitling Chronograph. Leon Breitling was a brilliant watch maker, but marketing was not his strong point. If Hans Wilsdorf and Rolex had accomplished this, you can bet he would have let the World know about it.

Leon Breitling *Circa 1909*

1914 Leon Breitling died at the age of 53. His son Gaston Breitling took over the watch company. He could see what his competition had achieved. So over the next 12 months he completely changed the company. He brought in new machinery and In 1915 produced Breitling first chronograph wristwatch. The watch was a favorite with pilots around the world.

WWI came fast and so did the aircraft industry. Gaston Breitling produced many instruments for the early bi and tri planes. The First World War and the German war machine proved to be a huge market for Breitling. Gaston Breitling gave many watches as gifts to High Ranking German officers. It allowed him to keep his skilled workforce, rather than send them to fight. As he was building instruments for the German Air Force it was allowed, Gaston could also continue to build watches. The exact connection between Gaston Breitling and the German war machine is believed to be only business, however many ask why the symbols are similar. They are similar, but as you read on you will learn that Breitling did also supply the allied air forces as well. There is absolutely no connection with Breitling and the German Military, SS or Nazi party. Breitling was a quality watch it was then, like it is today worn by the elite. In Germany in the early 1940's the SS believed that they where elite and wanted everything in this class including fine art, Mercedes cars, Breitling watches and Champagne. The next page has the Breitling symbols, but what you may not know is that the Breitling anchor and wings logo was not introduced until 1962. Long after WWII had finished. The Anchor with a large B and wings are to represent Breitling in the water, on the land and in the air.

At Breitling's beginnings the watches where marked on the underside of the dial. From the outside you could not tell it was a Breitling. The name at the time on the watches was 'Montbrillant'. In 1930 the watches became signed with *'Breitling'* Script. This remained until 1962. However the Navitimer has always used a logo with wings.

WW II.German navy anchor badge. *Circa 1929.*

Circa 1932

WW II German naval Officers lapel badge and the Nazi military lapel. *Circa 1932*

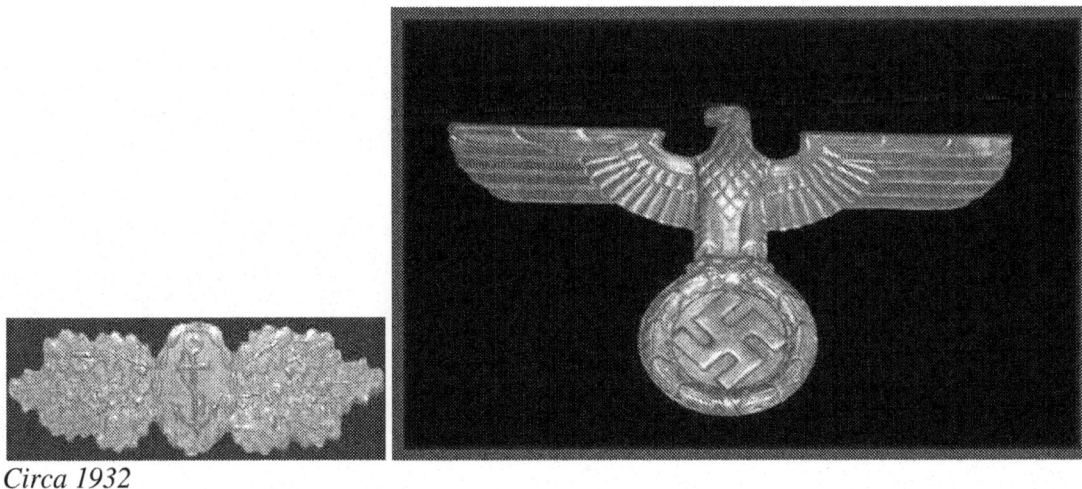

Circa 1932

German air ace Manfred Von Richthofen was a squadron leader for the German Luftwaffe. He reported to have shot down 76 allied planes mostly British. 60 of Manfred's kills where in his famous Foker Tri plane. Included in his victims was the British flyer ace Major Lanoe Hawker VC (Victoria Cross). Hawker was referred to as the British Boelcke. While Manfred Von Richthofen was known as 'The Red Baron'.

He wore a Breitling watch on all his flights. Manfred was shot and killed in a dog fight over Austria on April 21st 1918. Roy Brown of the British Royal Air force fired the fatal shot. Although Manfred still mannered to land his plane. He was captured by a sector managed by the Australian Flying Corps. His last word to his captives was "kaput" ("Broken").

In 1922 Sidney Australia a Breitling watch was sold at auction for $445 Australian dollars. In today's money $30,000 US. It was bought by an unknown bidder and sold as 'The Red Barons' watch he was wearing whilst shot down. Today the whereabouts of that actual watch is unknown. It is thought to be with a private collector in Japan.

Circa 1918

The 'Red Barons' famous Red Foker plane.

Breitling was important to both the first and second World war. Its instruments proved invaluable to pilots. It was mostly the German Luftwaffe that benefited. Adolf Hitler awarded some early Breitling watches to top ranking officers of the SS. Eva Braun who became Mrs. Eva Hitler 29th April 1945 was given a Ladies Gold Breitling Streamline Chronograph watch. The art deco designed watch with a tan leather strap was given as a wedding gift by her new Husband Adolf Hitler. The following day, after verifying the potency of a cyanide capsule, by giving their dog 'Blondl' a capsule. Eva and Adolf Hitler committed suicide.

Adolf Hitler and proud German sailors get ready to pose for a picture. This very rare picture was taken by a sailor. To Hitler's left Gunner Heinz Von Schwint admires his new Breitling Chronograph he had just been awarded along with the Iron cross, on board the German battleship Tirpitz. *Circa 1941*

Hanna Reitsch

Among the German Luftwaffe officers was Hanna Reitsch, She was given a Breitling watch by Adolf Hitler. He also personally awarded her the Iron Cross and the German Luftwaffe Diamond clasp. She was the highest decorated woman in Germany.

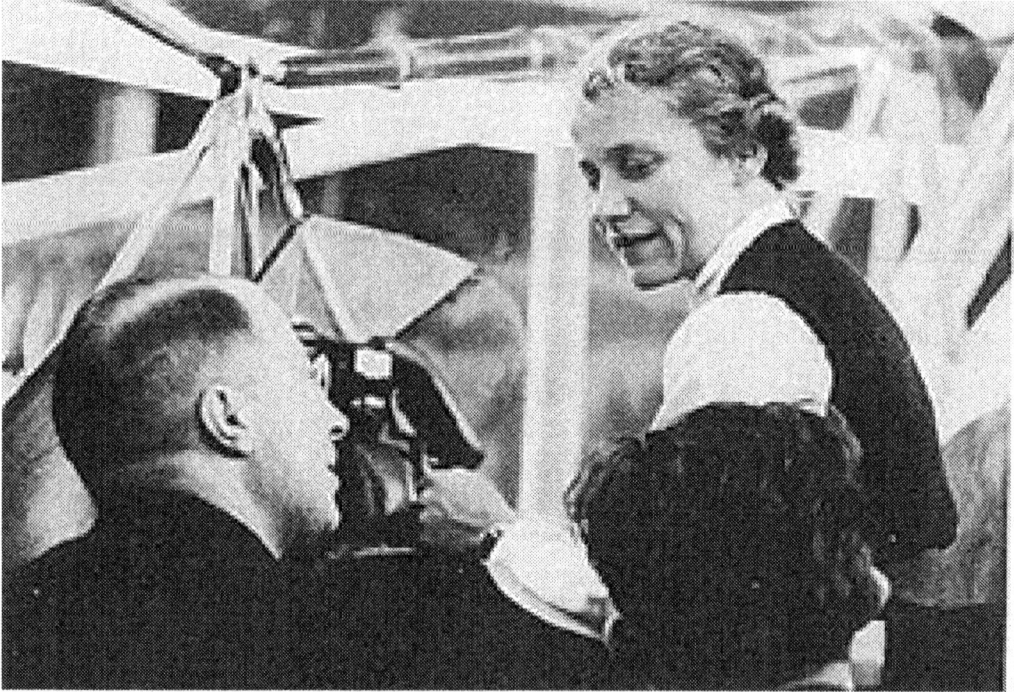
Circa 1936

Hanna Reitsch, Climbing on board the world's first Helicopter. (The Focke-Achgelis) On her left wrist, her Prized Breitling given to her by Adolf Hitler.
Circa 1936

Hanna Reitsch was also the first person in the world to fly a Rocket powered plane. And was also the first person to fly across the Swiss Alps in 1932.

She flew the V-1 Rocket and the rocket powered Messerschmitt 163. Hanna also set over 40 altitude and endurance records. She also flew the last plane out of Berlin in April 1945 minutes before the city fell at the end of World War II.

In 1961 she was honored at the White House by President John F Kennedy. At the White House, Hannah wore her Breitling watch.

She died in 1979 age 67. She was buried in Berlin with her personal effects that included her Breitling watch.

Above: Breitling pocket watch for German navy officer.

Above: German SS Major Helmut Kampfe pictured wearing his Breitling Chronomat.*Circa 1944*

34 Year old Major Helumt Kampfe SS Officer. Commanding officer of the 3rd Battalion Der Further Regiment. One of the highest decorated officers in the SS. Kampfe was awarded a Knights Cross & Given a Breitling Chonomat by Adolf Hitler. He was captured and killed by the French Resistance in 1944. In retaliation this sparked off one of the worst massacres of WWII. In the French Village Oradour-Sur-Glane June 10th 1944 over 400 woman and children were killed. In total the SS reported 643 Enemy deaths. Many believe it was far greater.

<u>Breitling Aircraft clocks</u>

Top picture: an early Breitling watch produced in 1918 for the German Foker plane. *Below:* an A-10A Aircraft clock from a United States 1950's C-54 transport plane. This was recovered after being shot down in Vietnam.

In 1923 Breitling produced the world's first independent chronograph push piece. Start and return-to- Zero functions had before been controlled by using the watches winding crown.

1927 Gaston Breitling died July 30th.

1932 Willy Breitling takes over control of Breitling,

1933 Breitling World Cup of Aerobatics was introduced.   Pilots fly free-style for 4 minutes and this is judged, by a World-wide panel of judges.

1934 Willy Breitling develops the second return –to –zero time push piece. Willy's new invention was far superior to the model produced 11 years earlier. It was now possible to measure several successive laps' times and with a push piece. This gives the Wrist chronograph its definitive form.

1935 Amelia Earhart wearing a Breitling watch flies solo across the Pacific from Honolulu to Oakland.

1935 Breitling water-proof Chronograph *Circa 1935*

1936 Breitling became the official supplier to the British Royal Air force (RAF). Breitling failed to launch an advertising campaign in the UK to coincide.

1942 Breitling launched the Chronomat, (See watch facts at end of book for Chronomat/Chronograph) this was the first chronograph to be fitted with a circular slide rule. Plus Breitling now became the official watch to the American armed forces. The watches where engraved with a logo; BREITLING WATCH

CORPORATION U.S PROPERTY.MILITARY.

During WWII Breitling supplied American, British and German airmen. After the war many of the serving pilots took jobs as commercial airline pilots and continued to wear Breitling watches. Today the Breitling is still a favorite of Pilots across the world.

Willy Breitling's Company was now producing 250 models. 212 of these were wrist watches. Breitling produced a catalogue for suppliers; this became known as the 'Chronolog'. The Chronolog is still produced today and comes out every November.

Chronographs are very complex; they are fitted with a variety of dials and can perform various tasks. The best feature of the Chronomat-Chrono-Matic is due to its movements. It has a basic movement and a Chrono-section unit. These 2 units are attached with locking screws, on top of each other. The winding crown is on the opposite side at the 9 hour. To wind the watch you would probably have to remove it from your wrist, but as it is an automatic movement this would not have to be done often, if ever. When the chronograph is started the swing drive is brought into contact with the second-counting wheel of the chronograph by means of a rocker. When the stopwatch is stopped, the swinging drives swings out and the second-counting wheel is blocked by a tiny brake.

1943 Breitling launched into the United States. The Breitling Watch Corporation of America was formed.

1952 Willy Breitling moved the company's head quarters to Geneva, although the production of the watches remained at La Chaux-De-Fonds. Breitling had grown so large that it chose to cease the production of watch movements. It relied on suppliers. This also allowed Breitling more room for design changes such as shape and size of the watches.

The Navitimer

Willy had also been working with his target customers; pilots. 1952 Breitling created the Breitling Navitimer. This was a watch equipped with the 'Navigation Computer' capable of handling all calculations called for by a flight plan. The Navitimer soon became a favorite of pilots and 'wanna be' pilots around the world. Breitling was already a major player in supplying major international airlines with cockpit counters. The Navitimer took the watch company to a new level, a level only Rolex could rival.

An early Breitling Navitimer *Circa 1952*

1957 Time magazine posted the Breitling Navitimer on the front cover, and had a 5 page article on the Breitling Navitimer. Due to copy right we are unable to produce the front cover here in this book. The magazine carried advertisements for the Breitling Navitimer. At the time it would sell for $95.

1962 Breitling made another first. Astronaut, Lt Commander Malcolm Scott Carpenter wore a Breitling chronograph on his orbital flight around the earth aboard the Auroa 7 space capsule. Despite this again Breitling never heavily publicized the fact he was wearing a Breitling. A few poster ads did appear, but nothing like what Rolex would do. Breitling still did not have the marketing skills of Rolex's Wilsdorf. Despite this the Breitling Cosmonaute was a success.
The watch passed all the extreme tests NASA could give it. Many of these tests where on the ground using very fast machines that would spin the watch at speeds of over 1500 miles an hour. The force of gravity would stop most watches; Breitling passed all tests and was therefore chosen for the first trip. From the information I can obtain, Rolex also passed the tests.

Watch in Space!

This watch orbited the earth in May 1962.
It is the NAVITIMER, the watch that timed the Astronauts,
produced by

Breitling

Geneva

Appointed Watchmaker to World Aviation

BREITLING

G.-Léon Breitling S.A., 6, Place du Molard, Geneva (Switzerland)

This advertisement is running in the international publications "Life" and "Newsweek"

Circa 1962

Below Astronaut Malcolm Scott Carpenter climbing into Auroa 7, wearing his Breitling.

Circa 1962

1969 Breitling invented the automatic (Self-winding) chronograph movement in partnership with movement manufactures Buren and Heuer-Leonidas. This was a major break though the entire Swiss watch industry. The same year Boeing takes off carrying 402 passengers with its B 747 known as the Jumbo jet. The same year Concord takes to the sky for the first time. Both planes where fitted with Breitling instruments.

1975 Breitling like many others produced its first quartz watch.

1976 Breitling produced Navitimer Quartz with LED display. The LED would automatically turn itself off after 3 second to conserve batteries.

Breitling Navitimer Quartz Movement. *Circa 1972*

1979 Willy Breitling died. His 2 sons Gregory and Alain Breitling were too young to take over the running of the company. It was Ernest Schneider. A former pilot and micro electronics expert who takes over the Breitling Company in April of this year. Willy Breitling died May of this year.

Ernest had previously worked for the Sicura Company. As well as being a pilot Ernest Schneider also enjoyed sailing. He launched a range of watches for skippers and divers. This was direct competition for the Rolex GMT and Submariner. Ernest kept the Breitling name and modernized the factories. He brought in new equipment. One thing Schneider did was to keep the quality and ensure that Breitling remained as one of the world's most precision watch manufactures. He consulted with airline pilots. He insisted that Breitling remain the pilot's choice of watch.

1980 Breitling produced the Breitling Jupiter, Breitling Mars and Breitling pluton. All three where Chronographs and designed with pilots in mind. Ernest Schneider had proved all his critics correct, he was just what Breitling needed to take the company to the 21st century. Ernest Schneider also brought the company up to date with marketing. Never again would Breitling fail to use the Worlds media for brand recognition. At last Breitling had its own Hans Wilsdorf.

1984 Breitling Re-launched the Chronomat; the watch was produced with a heavy bezel with rider tabs. This was a return of the Chronograph and soon became Breitling's top selling watch.

1985 Breitling produced the Breitling Aerospace. The new watch was a multi-functional electronic chronograph made in Titanium. It became a top seller among airline pilots.

Circa 1952

1990 The 'Emergency' watch was created by Breitling. This is a multi-functional watch The Emergency is an instrument watch with built-in micro transmitter broadcasting on the 121.5 MHz aircraft emergency frequency. Following a crash or a forced landing, the Emergency will broadcast a signal on which rescuers can home in. The Emergency's transmitter is activated by unscrewing a protective cap and pulling the antenna out fully, it will then broadcast for 48 hours. The transmitter's signal has a range of about 100 miles. The watch has been a huge success and favorite among solo pilots in light aircraft. Some mountaineers also wear the watch.

The Breitling 'Emergency' watch *Circa 1969*

1992 Breitling re-launched the Navitimer. Breitling under Ernest Schneider had produced the World's smallest movement; they developed with the Caliber 33. The movement was known as the Navitimer Airborne. The watch typically had 39 jewels and a large gold rotor.

1998 Breitling produce the B-1. The B-1 is the World's most versatile multifunction chronograph manufactured. It was designed in cooperation with aviation designers; it is fitted with a microprocessor specially developed for Breitling. The B-1 is the most advanced watch of its time; it embodies significant progress in the firled of Swiss micro-electronic engineering.

Breitling B-1 *Circa 1997*

1961 Breitling Chrono-matic *Circa 1960*

BREITLING
1884

INSTRUMENTS FOR PROFESSIONAL

Breitling Orbiter 3

Breitling has recently set a new record for aviation by being the first non-stop round the world flight in the Breitling Orbiter 3, the balloon flown by Bertrand Piccard and Brian Jones. The Breitling Orbiter 3, a hot air balloon featuring the most advanced technologies. Orbiter 3 took off from Château-d'Oex in Switzerland on March 1st, 1999 at 08H05 UTC, flew 45'755 km and landed on March 21st, 1999 at 05H52 UTC in Egypt completing a world breaking non-stop flight of 19 days 21 hours 47 minutes. The flight was made possible thanks to the proficiency of the crew and the efficiency of the entire Breitling team who worked on what can be considered the last major aeronautical event of this century.

Piccard and Jones wrote a book entitled "Around the World in 20 Days" about their adventure. The book was published in 1999 by John Wiley & Sons, Inc. The ISBN number is 0-471-37820-8. It is a great read for anyone interested in aviation or real life adventures. University Studies are making a film about the flight, filming starts in January 2008.

BENTLEY

Circa 2004

If you have read the entire book you will now know that Rolex got part of its name from Rolls Royce. Alfred Davis was an owner and admirer of the World most luxurious motor car. Rolls were purchased by the Germans and no longer British. HM Queen Elizabeth II now rides in a Bentley. The Bentley hand made in Great Britain soon became the World's most perspicuous motor car.

Bentley used only the best materials in its cars. In an age of digital clocks and watches the Bentley installs Breitling clocks.

The dashboard of a New Bentley *Circa 2007*

Breitling launch the Breitling Bentley watch. The new watch can only be described as HUGE. It comes in at 8 oz. That's half a pound of watch on your wrist. It measures 1.5 inches (43 mm) across the face.

The chronograph automatic watch despite its size is a fast seller. It was the top selling Breitling mans watch in 2007.

Breitling Fakes and Replicas

2 Breitling Bentley watches. *Circa 2005*

The top watch has a gold color metallic face. Both watches 8oz and well made. However the top watch is a poor fake, probably grade 2. The fake is also automatic, but open it and find a cheap mechanism and plastic ring to hold the workings. Plus none of the workings are marked; the fake did come in a box, with certificate.

Points to look for with Fake Breitling's.

1. On a Genuine Breitling the window for the date is large and eats into the second and minute markers, the fakes like above do not.

2. Original Breitling's do not have raised sub dials.

3. The inner bezel on the fake does not blend seamlessly into the dial; you can see a black ring on the picture above. The genuine watches do not have this.

4. Breitling and wings logo are embossed on genuine watches, on the fakes they are stamped on.

5. On a Breitling fighter watch the finish is completely brushed. On the fake it's a gloss finish.

6. The back case. The original has markings on the underside, Normally 4 -6 letters and number on the 2 closest links, the fakes do not.

7. Breitling watches have the crystal made from a scratchproof sapphire that is non-reflecting. This not only protects the dial against the suns UV rays, but removes optical disturbances. The fakes I have seen reflect light and the watch may have to be put at an angle to read the time.

The dead giveaway. On an Original Breitling, Look at the very center pin that holds the minute, hour and second hand. It should all be the same color, normally silver. The fakes all have a black pin. Please read through the points on Rolex fakes, many points are the same and must be used when buying a new watch.

Did you notice? The watch on the front cover? It's a fake Breitling Bentley!

Ernst Benz (Breitling)

If you have ever flown an aircraft you will know the name Ernst Benz. The famed German company that produces instruments for aircraft cock pits.

The company has the best reputation in the world. To capitalize on this Ernst Benz came up with the idea to produce high quality wrist watches. However the market was already taken by Breitling and Rolex. Ernst Benz negotiated with Breitling; the deal was not marketed heavily, as Ernst Benz already had a good reputation.

Pilot reading material (Plane & Pilot, Professional Pilot and Light-Sport Aircraft) Magazines where used to advertise the first watches produced by Breitling for Ernst Benz.The watches are high quality as you would expect and produced by Breitling. If you open the back case, you will find the Breitling markings on some of the Ernst watches and on the watches produced after 2000 they now all have Ernst Benz inscribed. Perpetual Chronographs to appeal to pilots are the main watches available.

Ernst Benz 47mm Stainless Steel Chronograph *Circa 1992*

The Breitling Stop Watches.
A Stop watch is a precision instrument that measures small amounts of time or to add up several small amounts of elapsed time segments. In other words they start and stop at the push of a button to record time.

For over 90 years Breitling has been world leader in stop watches. Today they can still be found under the name Breitling or Sprint. The original stop watches where also branded under Montbrillant name.

Circa 1979

The original Stopwatches where mechanical, today some are quart. The quartz stop watches are all produced under the brand name sprint. Although the Sprint brand has also been used to produce mechanical stopwatches.

By getting a foot in this market early, Breitling has remained number 1 in the world. Most athletes and trainers will own a Breitling stop watch. When 25 year old English man Roger Banister broke the world record 6th May 1954 to be the first man in the world to run 1 mile in under 4 minutes, this was recorded on 3 Breitling stop

watches. The time recorded was 3.54 minutes.
Circa 1970

Rolex and Breitling remain today the most accurate watches in the world. Official Chronometer Certification was awarded to Rolex in 1910 from the "Bureau Official" in Switzerland. This was the first time it was ever given to a wrist watch. Since then Rolex and secondly Breitling hold more records and certification than any other watch. Third place goes to Omega. Although Omega, Cartier and Tag Heauer continue to apply for certification, Rolex and Breitling each have more certificates than all the other manufactures put together. *Omega was started by 23 year old Louis Brandt 1897.*

Rolex and Breitling continue to be the World's top manufacture of high precision wrist watches. They continue to get awards for accuracy.

The first timing contests for watches were in 1869, and of course in Geneva at the Geneva Observatory. Throughout the years the tests were made more and more severe. The format of these competitions are part to organize chronometrical standards. Across the world such as Kew, London England, Yale in the US, Geneva and now in 2007 Tokyo Japan.

In the 1980's many people rejected mechanical watches and switched to quartz movements. Citizen produces the *Eco-Drive* system, these watches are solar powered and never need winding or batteries. However as we approached the 1990's we saw a rebirth of precision mechanical movements. Rolex and Breitling top the market in this area.

As we move into 2008 Rolex and Breitling have greater World-wide sales now than they have ever had. The only company coming close is Omega.

How can I contact Breitling?

In the US:

In Switzerland:

BREITLING USA INC.
HANGAR 7
206 DANBURY ROAD
WILTON, CT 06897
sales@breitlingusa.com
Tel: +1-800-641-7343
Tel: +1-203-762-1180
Fax: +1-203-762-1178

BREITLING SA
P.O. BOX 1132
2540 GRENCHEN - SUISSE
Tel: +41 32 654 54 54
Fax

Current Breitling Models:

Aerospace Advantage, Airwolf, Avenger, B-1, B-2, Bentley, Blackbird, Chronomat Evolution, Cockpit, Colt, Colt Oceane, Co-Pilot, Hercules, Navitimer, Skyracer, Starliner, Steelfish X Plus and Superocean

Following the huge success and record sales of the Flying B, Breitling continues to provide professional pilots with the latest technological advancements. This time the Breitling Company has enriched its Professional line with the new Airwolf Electronic chronograph offering a variety of new functions. The professionals in the field of aviation will no doubt appreciate the Airwolf's supreme performance, efficiency and distinctive design. The functions include: .Alarm, 1/100th of a second chronograph with split and additional times. Countdown timer, second time-zone with independent alarm, UTC (Coordinated Universal Time) perpetual calendar.

The case is made of Stainless steel. The Breitling Airwolf chronograph measures 43.5 mm in diameter. Protected by a glare-proofed sapphire crystal, the case and case back provide water-resistance to 50 meters.
The watch's face reveals two big-sized LCD 12/24-hour analog and digital displays, applied "3" and "9" numerals and hour-markers. Airwolf's legibility is provided with oversized hands and display backlighting usually featured by the NVG - night vision goggles. The decoration of the dial has been created to remind a jet engine - a kind of Breitling's tribute to aeronautical vocation. The bidirectional rotating bezel (slide-rule) is pinion and guarantees optimally precise read-off.

INSTRUMENTS FOR PROFESSIONALS
BREITLING has a single-minded commitment to building ultra-precise and ultra-reliable wrist instruments intended for the most demanding professionals. Our obsession is quality. Our goal is performance. Day after day, we consistently enhance the sturdiness and functionality of our chronographs. And we submit all our movements to the merciless testing procedures of the Swiss Official Chronometer Testing Institute (COSC). One simply does not become an aviation supplier by chance.

AVAILABLE FROM SELECTED JEWELERS THROUGHOUT GREAT BRITAIN AND IRELAND.
FOR YOUR NEAREST STOCKIST TELEPHONE 020 7637 5167.

WWW.BREITLING.COM

BREITLING
1884

INSTRUMENTS FOR PROFESSIONALS™

UK Advertisement. *Circa 2002*

109

The worlds other quality watches

Ball

The detailed movement of an early Ball pocket watch. *Circa 1928*

Ball diver wrist watch. Tritium hour markers and hands. *Circa 1992*

Ball watches are the only watch in this category manufactured in the United States.
Webb C. Ball was born in Fredericktown, Ohio on October 6, 1847.
When Standard Time was adopted in 1883, he was the first jeweler to use time
signals, bringing accurate time to Cleveland.
In Kipton, Ohio April 19th 1891, known as the Kipton disaster a mail train struck
another. The train's conductor's watch had stopped and restarted. Because of this
over 20 people were killed. Webb C Ball was enlisted in July of this same year to
investigate times and conditions for the railroads. His official title was inspector of
the lines. This inspection system appeared to be the beginning of the vast Ball
network that would encompass 75% of the railroads throughout the country and
cover at least 175,000 miles of railroad. Webb C. Ball also extended his system into
Mexico and Canada. He also introduced signals and produced watches for the
guards and drivers. A minimum of 17 jewels were specified Adjustments to
temperature and 5 positions were required. These 5 positions of adjustment were
dial-up, dial-down, pendant up, 9-up and 3-up. Later a sixth position was added,
pendant down. The watch had to be accurate in temperatures from 30 to 95

degrees. Ball watches were required to be lever-set, as a pendant set watch could pull out in the pocket and change the time. A clear white dial was specified with large numerals and 5-minute markers. Railroad watches were required to be inspected by a watch repair store or the railroads own time inspectors every 21 days and a tolerance of 30 seconds plus or minus, for the period of 2 days was required. He also produced some early anti magnetic watches, before Rolex did, however due to the cost; they never went into full production.

Ball was appointed president of the Hamilton watch company. (Today Hamilton is owned by the Swatch Watch Company)

Ball opened a jewelry store and called it the Ball Watch Company. Ball used movements form top watch companies such as Hamilton and Waltham. Webb ball died August 16th 1922.

In 1941 the company switched to Swiss movements.

1982 Ball Trainmaster wrist watch. *Circa 1975*

The company has been owned by the Ball family until 1991. Today the company still produces watches for sportsmen and railroads.

Today Ball is recognized as one of the World's premier watch companies and rightly deserves a place in this book.

2007 the watch company still keeps their Railroad connection. The current models are the Conductor, Engineer Hydrocarbon, Engineer II, Engineer master II, Inspector, Trainmaster and Fireman

Cartier

Ladies Cartier Tank Gold Tank watch. *Circa 1972*

Cartier watches started out back in 1847. Cartier is the World's oldest watch company. Louis-Francois Cartier the son of a Brass Horn (Trumpet) maker started the company in Paris France. Louis Cartier a goldsmith made Jewelry and supplied the worlds Royal Families. His work was elegant and very fine. Louis-Francois Cartier was born in 1875 and died in 1904.

In 1874 Louis passed the business to his son Alfred Cartier. Alfred started production of elegant watches. (Alfred Cartier 1841-1925) In 1899 Alfred passed the business to his son Louis Cartier. Louis Cartier started to produce more pocket watches; his aim was to specialize in watches.

In 1907 Louis Cartier signed an exclusive contract with watch movement manufacture Edmond Jaeger. Cartier expanded the business and open branches in London, England. New York and St Petersburg. The New York office was run by Alfred youngest son Jean-Jaques Cartier. The second Son Pierre Cartier ran the London office.

Cartier watches soon gained a reputation of very high quality and precision. In the early 1920s Cartier formed a joint company with Edward Jaeger (from the Jaeger-Le Coultre company) to produce movements exclusively for Cartier. Thus was the famous European watch & clock company was born, although Cartier continued to use movements from other makers. Cartier watches can be found with movements from Vacheron Constantin, Audemars-Piguet, Movado and Le-Coultre. It was also during this period that Cartier began adding its own reference numbers to the watches it sold, usually by stamping a four-digit code on the underside of a lug. In fact, many collectors refuse to accept a Cartier as original, unless these numbers are present. However the numbers are just stamped on and could easily be copies, I would not accept just this as prove that the watch is genuine.

In 1931 Cartier designed its first waterproof watch. This had a 3-piece case and was manufactured in France.

1933 was the year when Cartier was commissioned by the Pasha of Marrakech to create a watch he could wear while swimming. As one of the earliest truly water-resistant designs, the watch featured an attractive round case of solid gold.

1941 Cartier produced the 'Moonphase'. During the 1940's and for the first time in Cartier history, Stainless steel was used. This was due to Gold being less available during the war.

1942 Louis & Jean-Jaques Cartier died. The company still in operation slowly started to go downhill. 30 years later a group of investors took over the company and appointed Alain Perrin as CEO. Perrin a former and very successful antique dealer introduced many new lines and quickly turned the company fortunes around. Cartier's best selling watch is the highly fashionable dress watch, 'The Tank', seen over. Today this is still a best seller for the company. Throughout the years Cartier have use movements produced by Edmond Jaeger, Movado, Le-Coultre and Audemars-Piguet

By 1968 Cartier had evolved from a family business into a huge multinational company.

1972 Joseph Kanoui headed the financial syndicate which bought control of Cartier Paris. Robert Hocq became president of the company. He once again united the three branches of Cartier and took over the London and New York Management in a move to re-establish Cartier's image of prestige and importance.

Out of all of the watches produced in 2001-2007 by Cartier 75% are the ladies watches. The jewels and designs keep Cartier ahead of competition in this area. Needless to say, a Cartier watch is finished to very high standards. The cases and bracelets in particular are meticulously handcrafted and exude quality in every sense of the word. Yet in spite of the famous brand name and timeless designs, Cartier watches are available in a wide range of prices. You can get a new Cartier for under $1500.

In recent years we have seen a sharp decline in Cartier watch sales; it could be a phase that they and the rest of French producers are going through. Since the French opposition to the invasion of Iraq, Many Americans and British have taken their money elsewhere. The Cartier future is unknown.

The current Cartier collection include Tank S, Tank Americaine, Tank Francaise, Pasha de Cartier, Santos de Cartier, Roadster, Ligne 21, Panthere de Cartier, Tortue, Baignoire and Lanieres

The Santos watch was created in 1904. It was actually named after a good friend of Cartier, Albert Santos Dumont. Santos-Dumont was also an aviation pioneer, and his desire for a good timepiece to use while flying inspired the Santos watch.

World War I was the actual inspiration for Cartier's famous Tank watch. Cartier was both fascinated and impressed with the American tank. Thus, he proceeded to design a sturdy, yet lovely, watch called the Tank. A newer design, the Tank watch remains a popular style today. Its sharp-edged, vertical side-pieces and sleek lines echo the sturdy and simple construction of the American tank.

The Panthere watch (takes its cue from the elegant Santos watch, with its simple and supple links. The Panthere watch, however, introduced an extra-flat gold or silver link, which has given it a style all its own. The Panthere is at home as every day wear or for special occasions.

The Cartier Pasha was a man's watch until the 1980's; the Pasha watch was soon considered a much-sought after fashion accessory among women. The Pasha actually dates back to the early 1930's, when its conception was thought to have been brought about by the Pasha of Marrakech's desire for a waterproof watch. He asked Cartier to design a watch that could be worn as he enjoyed laps in his pool. This watch was actually a Tank watch, but by 1943, the Pasha watch was introduced. This watch features a large, round face, and is elegant yet solidly constructed.

Another one of the famous Tank watches, the Cartier Divan watch was introduced in 2002. It took the somewhat common rectangle shaped face and turned it on its side for a distinctive, yet classic, appearance. Considered a watch of the future, its innovative design and uncommon energy give watch lovers an exciting, new choice in the watch world.

Cartier watches offer a timeless and elegant design that is simple, geometric, and beautiful. Whether you are searching for a dress watch or every day timepiece, Cartier watches offer watch-lovers both options. Their quality is world-renowned.

A888
Chrono-graph 2002. *Circa 2003*

IWC (*International Watch Company*).

IWC Big Pilot Watch. *Circa 1990*

In 1868 Florentine Jones, a 27-year old American master watchmaker from Boston Started the IWC Watch Company along the Rhine River at Schaffhausen, Switzerland. The Company was first named IWC Schaffhausen Watch Company. Jones had the idea of uniting American expertise in automation with the legendary precision of the Swiss. He wasted no time in installing the machine tools most of them imported from America at premises in Schaffhausen. The factory pictured over was rented J.H Moser; the factory was previously used for producing hydro electric tools and machinery.

His pocket watches were of exceptional quality. The one floor in his plan was the huge import duty demanded by the United States government. Another floor in his plan is that the workforce in the Geneva region and remote valleys of the Jura Mountains, Feared Jones would eventfully take the production back to the United

States.

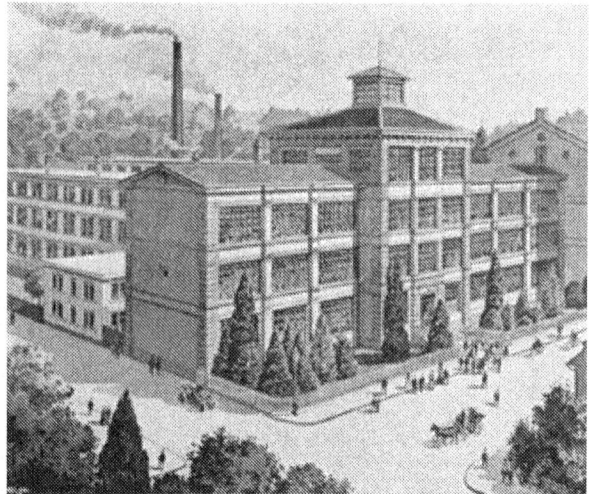

Circa 1899

Jones had to pay a premium for his work force.
Located in Schaffhausen, Switzerland, IWC Schaffhausen is notable for being the only major Swiss watch factory located in eastern Switzerland, as the majority of the well-known Swiss watch manufacturers are located in western Switzerland IWC did manufacture the World's first digital watch in 1885. (Digital meaning numbers on dials not LED display)
In the early 1900, IWC completely ignored the American market and concentrated on the British and German markets.

1927 Wrist Watch *Circa 1927*   1899 IWC Wrist watch. *Circa 1900*

*One of the oldest men's
wrist watches still working today.*

1936 IWC produced its first pilots watch with rotating bezel.
It was not until 1942 that IWC made its first wrist watches.

In 1944 IWC produced a range of watches called WWW. This stood for Watch, Wrist and Waterproof. It was produced for the British Army. On April 1st this year the factory in Schaffhausen was bombed in error by the United States Air Force, *Oops*! The factory was mistaken as a weapons factory.
However the bomb did not explode and fortunately no injuries were recorded.

Circa1980

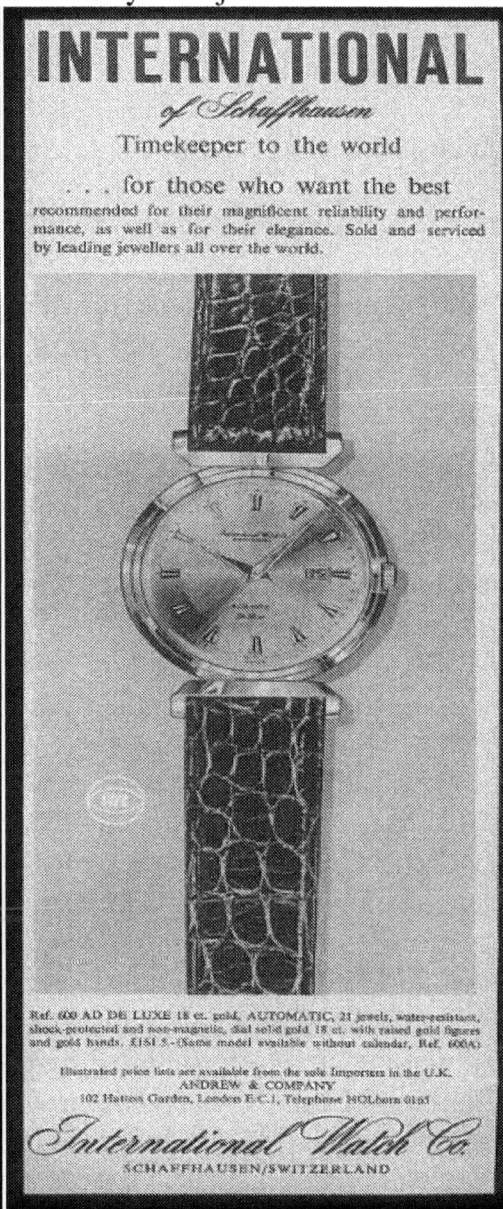

Circa 1962

1970 it produced its first quartz watch. Known as the 'Da Vinci' it featured the Beta 21 movement. This was a time when low cost watches produced in Japan flooded worldwide markets.

1978 The World's first wrist watch with a built in compass was manufactured in partnership with F A Porsche.

1980 The World's first chronograph titanium case is produced; again design by F A Porsche.

1985: The Da Vinci from IWC is the first chronograph with a perpetual calendar programmed for the next 500 years. It has the most accurate moon phase display ever as well as a four digit display showing the year in full.

1987: IWC presents the first rectangular, water resistant, automatic watch with a perpetual calendar - the Novecento.

1993: The Company celebrates its 125th anniversary.

2000: IWC develop the company's own movement for large wristwatches. The extra large 5000 caliber movement runs for 7 days and features a power reserve display with a Pellaton automatic winding system. IWC is taken over by Richemont.

2004: The Aquatimer range of watches is launched. The Portuguese family is extended to include the Portuguese Tourbillon Mystere, the Portuguese Minute Repeater Squelette and the Portuguese Automatic. New models are also added to the Da Vinci and Portofino lines.
Most modern movements in watches of IWC are not produced in house by IWC. Almost all movements are based on movements supplied by ETA. IWC also uses a JLC meca-quartz movement in their Portofino chronographs.

2007, IWC manufacturers the world's most sophisticated bracelet system, which requires neither screws nor pin and bushings to hold the bracelet together. Instead a solid pin is held in each bracelet link by a push-button lock on the underside of each bracelet link - allowing the pin to be totally locked in regardless of any damage that would normally dislodge traditional pin systems. Together with other beautifully manufactured mechanical gadgets like mechanical depth gauges, 7 day power reserve automatic movements, and deep-sea (6100 Ft (2000 meters) water resistant) resistant turning crowns for internal bezels, makes IWC truly a watch manufacturer for the future yet hand-in-hand with traditional hand craftsmanship.
IWC is most famous for its Flieger line of watches (This translates to Pilot in German) whose design date back to World War 2, the beautiful and traditionally made Portugieser line, and has just released a new range of highly engineered sports watches with many new inventions and patents called the Aquatimer line.
One of IWC's most sought after watch is the Mark XV, which traces back to WWII. Today this watch will fetch over $150,000 at auction.

Longines.

2007 Longines wrist watch
Circa 2001

1923 Longines White Gold pocket watch. *Circa 1923*

Movement of early Longines pocket watch *circa 1841*

Longines watch company started in 1832 by Swiss born watch maker Auguste Agassiz. The first watches and clocks where simply called Agassiz.

Auguste's first workshop was in St Imers Switzerland. His main workforce was home workers. He employed local people, just like the Breitling family. This was also the same village that Leon Breitling started his company. It is unknown today if Breitling did work on Longines watches, but thought highly probable.

The company was passed to his nephew Ernest Francillon in the summer of 1857. It was Francillon who built a factory in les Longines just outside St Imers. Most of the watch making was done in house apart from 20% by the outworkers. The watches now had the name Longines written on the dial.

Circa 1866

In 1879 Longines started to produce chronographs. They later produced aviator clocks and cockpit instruments. This is what is believed the very young Leon Breitling would have been producing at his parent's home.

1871 Longines started to sell to the American market.

Circa 1900

In 1889, Longines produced five different chronometers and gave to the Italian explorer Luigi Amedeo, who made an attempt to reach the North Pole. His failed attempt is scarcely noted in the history books. It was a huge disappointment for Longines, who had hoped for worldwide attention. 10 years later Robert E Peary was the first man to reach the North Pole. Peary was part of the U.S navy expedition.

1890 15th May, Longines registered the 'Wings Hourglass' trademark.

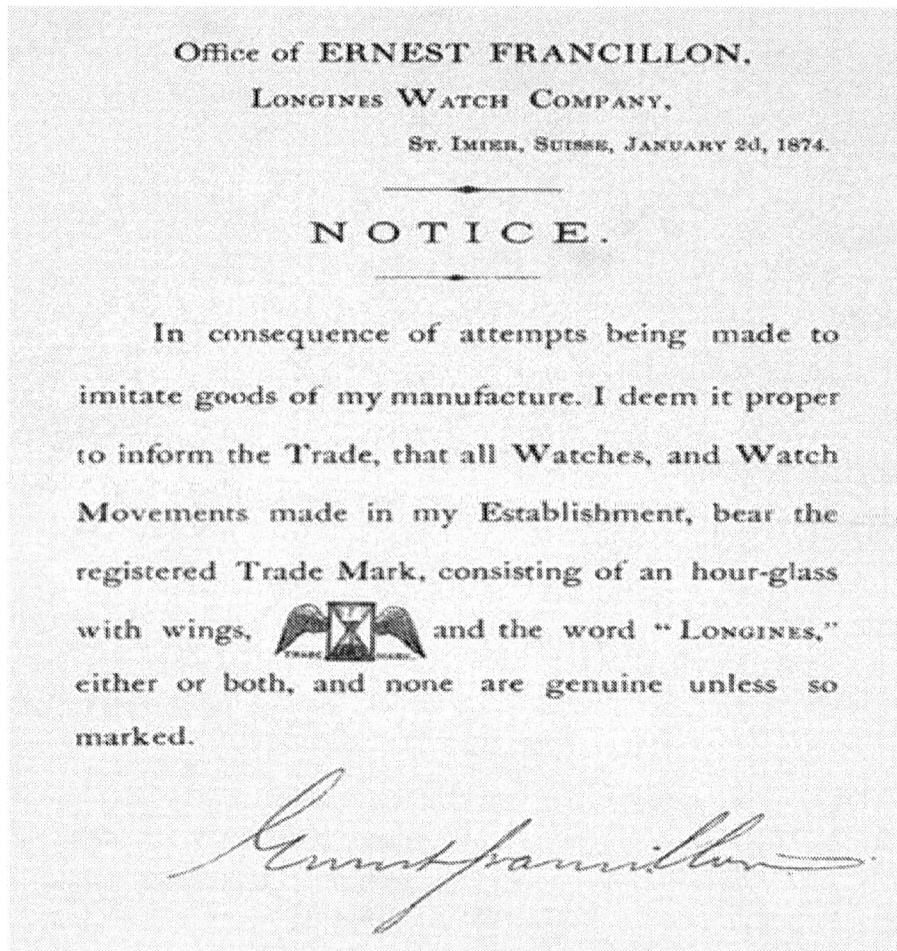

Circa 1911

1912 Longines produced the World's first automatic time keeping precision instrument.

1927 Charles Lindberg made the first nonstop solo transatlantic flight from New York to Paris in 33 hours and 30 minutes. The event was timed by Longines, Lindberg carried a Longines pocket watch and the aircraft was fitted with new instruments designed by Lindberg and manufactured by Longines.

1930 American born Richard Byrd was the first man to fly across the South Pole. He accomplished this in 18 hours wearing a Longines wrist watch.

1950 Longines purchased the Wittnauer watch company. The company was now registered as Longines Wittnauer. Watches are produced with both names on the

dials and some models with just one or the other name.
Since 1952 Longines have been the official time keeper for the Olympic Games.

1979 Longines broke a World record for producing the slimmest watch. It measures 1.98mm.
1994 Longines and Wittnauer parted company. The company was taken over by Swiss watch company SMH. SMH also own Omega and Swatch.

1995 Longines become official time keeper of the World games.

2004 the Olympic Games collection was marketed by Longines. The collection was for new fresh designs of Chronographs.

2006-2007 Longines become official time keeper for the Alpine World Ski Cup.

Longines is the World's oldest registered watch brand names. Today it is one of the most collected watches. An early 1920's watch that had an original cost of $8 will be worth today in excess of $15,000.

Omega

OMEGA is the last letter of the Greek alphabet and symbolizes accomplishment and perfection.

Omega 2006 Seamaster *circa 2001*   Omega 2007 Speed master *circa 20006*

This book is the history of Rolex and Breitling. I wanted to include these other brands; However Omega is not really just another brand. It must take third place in the world of watches.

Rolex was once worn by various actors' playing James Bond 007. Breitling has also been featured in two of the films. 2007 the latest James Bond adventure 'Casino Royale" has seen Bond (Daniel Craig) wearing and mentioning his Omega Speedmaster, *shown above.*

Many Omega watch collectors feel that now is their time and Omega may take a number 1 spot. Breitling may come under pressure from Omega but Rolex far exceeds the 2 companies in sales put together, nether the less, Omega is a truly magnificent watch brand.

23 year old Louis Brandt started the company in 1848 in La Chaux-de-Fonds. He started assembling watches and purchased his cases and movements from various

companies in Geneva and the United Kingdom.

Luis died in 1879 and left the company to his 2 sons Louis-Paul Brandt and Cesar Brandt. The company grew at a steady pace with no innovations.

In 1900 Omega produced a wrist watch; it was not only the first wrist watch made by Omega but also the Worlds' first wrist watch that was mass produced.

In 1903 both Louis –Paul and Ceaser died. At this time they where one of Switzerland's largest watch companies. It employed just over 1000 staff and produced over 250,000 watches per year. The company was taken over by the children of Cesar and Louis-Paul. The oldest was the 22 year old Paul-Emile Brandt. Paul-Emile worked to grow the company, however in 1929 with the War and difficult trading condition, made worse by shortages of material. Paul- Emile sold the company to the SSIH group. Omega was coupled with the Tissot watch company.

1916 Omega produced a wrist watch for the American troops who joined the allied forces. The watch had a steel grid to protect the crystal.

1917 British Flying Corps decided to choose Omega watches as there official time pieces. The US army did the same the following year.

1922 Omega Regulateur wrist watch was produced. This was a HUGE wrist watch; the movement 5970756 was previously used in a pocket watch.

(The dial across is a huge 55mm 2 ¼ inch)

1957 Omega Seamaster was produced. NASA decided to choose the Omega Speedmaster Professional as its official time keeper.

Omega Regulateur *Circa 1953*

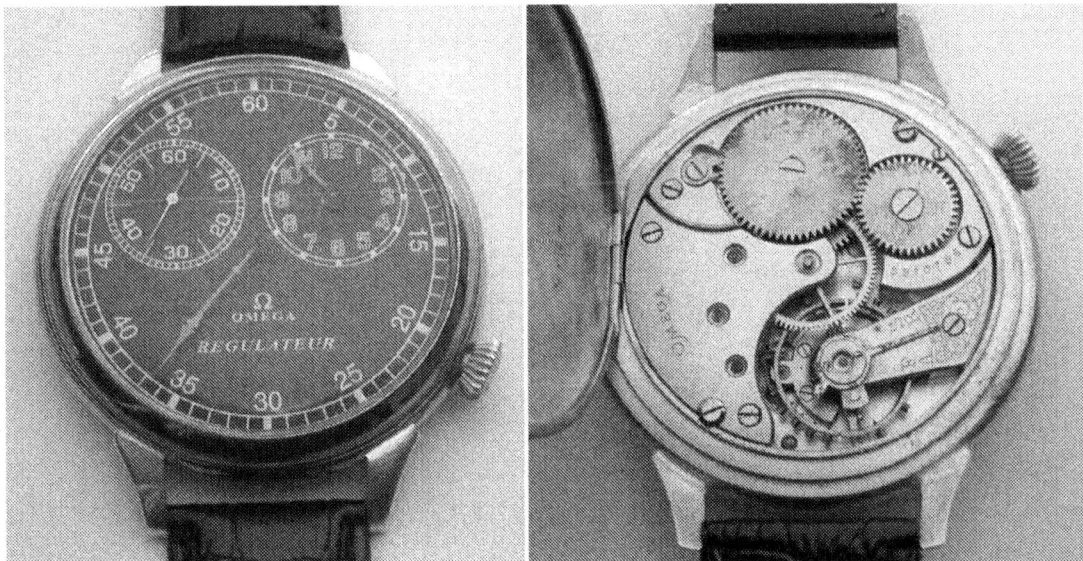

1969 21st July. American Astronaut Neil Armstrong became the first man to step on the moon. As he made the famous steps quoting "one small step for man, one giant leap for mankind", he was wearing an Omega Speedmaster chronograph.

1972, Omega received their two-millionth chronometer certificate.

By the early 1970's SSIH had become the World's 3rd largest watch manufacture. 1977 the year Elvis Presley died. (Elvis wore a Gold and Diamond Omega watch on his 'Aloha from Hawaii' world satellite concert). SSIH was in financial trouble. The company got a break from 2 large banks in September 0f 1977, but the future looked bleak. Interest rates increased in the early 1980's putting pressure on the company and the banks.

On the 18th of May, 1983, Omega received its 100,000th official rating certificate for quartz chronometers, also in 1981 Seiko put an offer on the table to buy the troubled company, but this was refused. The banks who had lent the money questioned the decision, but as long the monthly payments were made SSIH was still in control.

1985 SSIH merged with ASUAG (Longines) and formed the new company SMH. *Société suisse de Microélectronique et d'Horlogerie.*
In 1998 the name was changed to the Swatch group. In the group Omega remains the flagship brand name.

Today Omega remains one of the World's top watch manufactures and is worn by celebrities such as Tom Cruise, Nicole Kidman, Cindy Crawford and Pierce Brosnan.

Cindy Crawford's Choice

Pierce Brosnan's Choice

Ω OMEGA   HALTOM'S

Ω OMEGA
my choice

Circa 1998

The current 2007 Omega collection include the Omega Constellation, Seamaster, Seamaster GMT, Chrono Diver, Diver Chronometer, Racing Chronometer, Aqua Terra Chronometer, Apnea, Aqua Terra Railmaster, Omega X-33, New Omega Speedmaster Professional, Moon Phase, Broad Arrow, Legend, and De Ville Co-Axial Power Reserve.

As with all top dollar watches there are many fakes, replicas and aftermarket Omega watches available. As they are not such a common watch it is harder to spot them. Follow the steps on how to avoid buying fakes on the Rolex and Breitling section. If a deal seems too good to be true, well it probably is.

Caveat Emptor, Buyer beware.

The Rare Omega Automatic Coaxial Central Tourbillon

Omega developed the Automatic Co Axial Central Tourbillon to celebrate its 100th anniversary in 1994. The 18k rose gold case has dimensions of 39mm diameter by 11.7mm thick. The case looks and feels massive and sturdy, yet with its design embellishments, like the flared lugs, there is also an unmistakable elegance to its lines. As you would expect on such a watch, the fit and finish are outstanding. The case is polished on all surfaces but the center of the case back which is brushed.

Technically, this watch is a great achievement. It's the world's first automatic central Tourbillon escapement. The hours and minutes hands are painted on two separate sapphire disks which revolve to display the time. The gearing for the sapphire disks has been moved to the outermost edge of the caliber to make room for the central escapement with its revolving one minute Tourbillon. The winding rotor is platinum for winding efficiency. The watch features two crowns, one in the traditional position for winding the watch when necessary and a second crown on the case back for setting the time. The caliber number is 1170.

The watch hit the US market in January 1995 with a price tag of $79,000. Sold in very limited numbers these watches today fetch $175,000 at auction.

Omega Automatic Coaxial Central Tourbillon Gold men's wrist watch. *Circa 1995*

Patek Philippe

18ct Gold Patek Philippe perpetual calendar chronograph *circa 2002*

Polish born Norbert de Patek moved to Geneva in July 1830. Geneva was then as it is now the World's capital of watch making. He worked in a variety of factories, 3 years later Patek and fellow Polish born François Czapek founded a small watch company. They called the small company Patek, Czapek & C°. They had the office in Geneva.

The partnership was short lived and in 1936 they parted company. Patek paid off Czapek, However with his partner gone and now all his company funds exhausted he desperately needed another partner. French engineer Adrien Philippe bought into the company.

At the Paris Exhibition of the Products of Industry in 1844, Philippe presented his latest invention, the first keyless winding and setting system that was viable on an industrial level. This was a giant step towards the modern watch, up until then all timepieces had to be wound and set with a separate key, which could be lost or incorrectly used. This ingenious discovery, which apparently left the Parisian watchmakers cold, immediately interested Antoine Patek... to the extent that the two men formed a partnership and founded the firm of Patek Philippe.

HM Queen Victoria acquired one of Patek Philippe's first keyless watches. Once the Queen of England had bought one, many rich and famous wanted the same. The result was a huge increase in sales for Patek Philippe.

1851 Patek Philippe took across the Atlantic to hit the huge American market. It signed an exclusive deal with Tiffany & Co. New York. Since then the company has grown at the same pace as Rolex and Breitling.

Like Cartier, the bestselling brand for Patek Philippe is the ladies watches.

Patek Philippe makes some of the most complicated watches today.

The caliber 89 in Gold, a pocket watch remains one of the most complicated watches ever made. It has two dials and 33 complications and comprises of 1728 pieces, 24 hands and eight discs. The watch was first made in 1980; it took 9 years to produce. It weighs 2.4 pounds. (1.2 kilo's)That's the heaviest watch made today. I would suggest you wear a strong belt to protect your modesty.

In the 2007 catalogue they have over 100 models and designs. They have the highest priced watches in the world. Fakes are everywhere, due to the high cost; it is recommended that you only buy from an authorized dealer.

Circa 2004

The above Patek Philippe watch has a price tag of $100,000. It's a 2005 Rose Gold Asymmetrical model and only 50 were produced.

Personally I can't see the attraction, but they sold out in 7 weeks when first put on sale. Today the President of Henri Stern Watch Agency, the U.S. division of Patek Philippe, is Larry J. Pettinell.

The current Patek Philippe collection includes Grand Complications, Calatrava, Gondolo, Golden Ellipse, Aquanaut, Nautilus and the Twenty 4.

Zenith. *El Primero*

Circa 1999

Men's Zenith El Primero Quantieme Rose Gold Watch

Last but by no means least Zenith.

The Company was found in 1865 by the youngest of all the previous company founders.

It was 22 year old Georges Favre-Jacot started his company and was a true pioneer in manufacturing of watches. Before this time most watch producing had been done by small watch makers and out workers or home/cottage workers as they came to be known. The watch company was simply called Favre-Jacot.
Jacot rented a large building that was previously used as an animal food store.
He started manufacturing watches; his aim was to build quantity but high quality.

1875 within 10 years he employed over 1000 people. The company produced pocket watches and dials for the British Royal Navy.

1911 Zenith is born. (Zenith is the name for the highest point in the universe). Jacot registered the name Zenith and used a star as its trademark symbol.
As from July of this year all the watches now carried the Zenith name.
By the early 1920's Zenith had sold over 2 million watches and had offices in London, Paris, Geneva, Moscow and New York.

He worked until age 82, when he retired. By the time he had retired in 1929 Zenith was one of the World's largest watch manufactures.

Military German Zenith Watch 1934 *Circa 1933*

Military German Zenith Watch 1934 *Circa 1933*

Zenith Pocket Watch, 1930's, stainless steel, case diameter 47mm, silver stick and Arabic dial with blue hands, sub-seconds dial; mechanical wind. *Circa 1929*

1959 Newspaper adverts *Circa 1959*

1969 Zenith introduced the El Primero; this was the World's first automatic chronograph movement. Its mechanical movement throbbed at a rate of 36,000 vibrations per hour. Today it is known by many as the smallest and most accurate watch in the world. Also this year Zenith and Movado join forces and produce a new range of men's watches.

1970, no surprises Zenith also produced quartz watches.
1980 Zenith re-launch the El Primero.
1991 Zenith produce the movement for the **Rolex** Daytona watches.
1999 Louis Vuitton buy the company. The aim to turn the company into the World's highest quality watch manufacture. *I can't see that happening!*

2003 The El Primero Open Chronograph watch is produced.

Zenith El Primero 'open' Chronograph watches. *Circa 2002*

Caring for your watch.

Rolex and Breitling watches are beautiful precision instruments; they are synchronized works of art. Comprising of literally hundreds of precision parts all working in unison to measure time, days and dates. A balance wheel on a movement vibrates backwards and forwards at a fantastic rate from 18,000-32,000 beats per hour. This is basically the heart of the watch. Just like us a healthy heart gives a healthy body. Many balance wheels have a tiny delicate hairspring that is welded by laser. It's this hairspring that allows a wrist watch to vibrate at such speeds. If just one part of a watch is 'out of sync' it will affect the accurateness of the watch.

I would recommend that you have your watch serviced every 4-5 years. The balance wheel on a watch will travel over 20,000 miles over the same time period. I would also suggest that you only use a Rolex trained repair man/woman for Rolex and a Breitling trained Technician for Breitling watches. This ensures that they have the correct knowledge and will also use genuine replacement parts if required. They will clean and use the correct lubricants.

You can't just take off the back case and give it a quick squirt of WD40.

Rolex and Breitling trained technicians will also clean and polish your watch and bracelet, it will come back looking new.

After this it is tested over a number of days, at different angles to ensure accuracy. A point to remember also is that ONLY an authorized Rolex jeweler or Breitling jeweler is authorized to provide you with a Rolex/Breitling warranty that is backed by the network or dealers.

Failure to get your watch serviced can result in dried out seals and gaskets, wearing on small parts and possible moisture damage. Once you get moisture in your watch, it can cause huge problems, the watch may have to undergo a full repair and replacement of many parts. The cost can be as high as $2000.

To locate an authorized dealer for your watch service, or go to the internet and search.

In the United States at Saint Paul University in Minnesota, Students are given a two year course on watch making programs, Rolex and Breitling fund a large part of this. The Graduated candidates are then given jobs across the United States with very lucrative benefits.

A question I get asked a few times is how does the Rolex Sea-Dweller compare to the Breitling Super Ocean Professional

The biggest difference between these two models is perhaps the price, although the Rolex no doubt is the true classic. The Rolex costs almost twice as much as the Breitling. However people often compare these two models when looking for a top quality diver's watch.

The Breitling Super Ocean is water resistant to 1,500 meters, (5000ft) and the Rolex Sea Dweller is rated at 1,220 meters, (4000 ft) both have the automatic helium escape valve. The Breitling's domed sapphire crystal is anti-glare coated on both sides while the Rolex has a flat sapphire crystal without coating. The Breitling is 41.00 mm in diameter and 14.99 mm high. The diameter of the Rolex is 40 mm and its height is 14.99 mm. Both the Breitling and the Rolex are available in matt finished stainless steel only.

The Rolex has the upper hand as far as the movements go. It uses the its own movement the caliber 3135 which no doubt commands greater respect than the out-sourced ETA 2824 movement used in the Breitling. That is not to say that the finely finished ETA will not last a lifetime; with proper servicing it very well may. The bracelet on both watches are of perfect for the job, However the Breitling is a bit bulky and can catch when putting on a wetsuit, where as the Rolex is slimmer. Few, if any, watches can compete with what many would call the ultimate dive watch -- the Sea-Dweller. However, the Breitling is an exceptional dive watch which will serve you well over a lifetime of extreme activities. And it is a fantastic value in comparison. In your daily diving life, either watch will serve you equally well. It's down to your budget and what you feel is the better looking watch. Personally I feel the movement used in the Rolex, that Rolex produces itself makes it a better watch and well worth the extra money.

The Breitling Super Ocean *circa 1981*   The Rolex Sea Dweller *circa 1980*

What you should know:

Always:

Keep your watch away from magnetic fields. If it gets close to a mechanical watch this can cause irreparable damage.
Clean your outer watch using a lightly damp cloth, only soap and water, never use chemicals.
Keep your Crown screwed in at all times. All Rolex watches are waterproof to 300 ft, but with the crown screwed in.
Wear your perpetual watch as much as possible, to keep it wound.
Get your watches serviced every 4-5 years, this is an expense we have to swallow.

Don't:

Polish the links on a Stainless steel Bracelet to make it look like something it's not, this will tarnish.
Over wind your watch.
Open the case unless it is really required.
Try to service the watch yourself.
Over-tighten the crown. You don't need to. The seal inside the crown has been designed to sit on top of the stem and form a perfect hermetic seal against all natural elements. The more you tighten it, the greater the pressure exerted on the seal which will eventually break with the pressure against the top of the steel stem... eventually losing its efficacy as a waterproof seal.

And finally;

Don't wear short-sleeve shirts in the middle of winter to show-off your Rolex or Breitling. You'll catch pneumonia.

Glossary of Terminology

Arabic Numerals: Numbers or figures on the dial. Representing the hours as opposed to Roman numerals.

Ardilon: This is the mobile often pointed part of a buckle which pierces the leather strap, when the buckle is fastened.

Automatic Winding: (self winding) the automatic watch is wound by the wearers movement. Washing shaving or brushing your hair with your watch on your wrist is enough to cause the weight inside the watch to rotate back and forth. The weight is connected by a gear to a barrel arbor, this is hooked to the watches mainspring, and it's this that does the winding and keeping it at a state of constant tension. See also perpetual movement.

Balance Spring: This is also called a hair spring. The hair spring controls the swing and movement of the balance.

Bezel: The metal ring around the crystal. See also rotating bezel.

Bubbleback: This is the term used to describe early Rolex perpetual. Due to the thickness of the case used to house the rotor movement.

Calendar watch: A watch with a mechanism that shows the date and sometimes month, Moon phase, and even the year. Most calendar watches have to be adjusted manually at the end of most months. However the Rolex perpetual calendar watch does this automatically.

Center Seconds: Also called sweep seconds, Mounted on the center post of the watch for much greater visibility, thus it makes it easy to read.

Chronograph: A watch or precision instrument that in addition to normal time telling functions, can also perform a separate time measuring function, such as the stopwatch, This has a second hand that can be started and stopped independently from the main time telling movement. Usually with buttons on the side if the watch, Look at Breitling Bentley. Chronographs have been very successful for Breitling.

Chronometer: This is spelt Chronometre in the UK. A Chronometer is a highly-precision watch which after very rigorous tests, has received a timing certificate from the official Swiss timing bureau. In plain English a Chronometer is certified to be very accurate. In blue collar worker British language, 'It keeps bloody good time".

Comex: The French company that produces commercial diving equipment. Comex stands for Compagine Maritime d' Expertise. The company is had part in the design of the Rolex one-way gas escape valve fitted to the Rolex sea Dweller.

Coronet: also known as the Rolex crown, this is the symbol of Rolex.

Cosmograph: Rolex trademarked term. For the Rolex Cosmograph watch.

Cosmonaut: Breitling trademark term. For the Breitling Cosmonaut watch.

Crown: The button on the side of the side to wind the watch or adjust the time, day/date.

Crystal: The glass that covers the watch face. Today many are Sapphire Crystals produced in laboratories.

Cyclops: The glass bubble fitted on some Rolex watch crystals. This is directly over the date. It will magnify 2.5 times. Rolex have the patent and first patent it in 1952. Look at the Rolex Datejust.

Dial: This is the face of the watch. The dial has the hour and minute markings on.

Duograph: Breitling Chronograph with sweeping second hand stopped by the crown.

Equation of time: The difference between real solar time and mean solar time, resulting from the Earths erratic orbit. In a typical year, the equation of time varies by about 15 minutes. Watches featuring an equation of time function are equipped with a clever mechanism to measure this 15 minute difference.

Gas Escape valve: A one-way valve used on the Rolex divers watches, Sea –Dweller and Sub-Mariner.

Gilt: Gold plated.

GMT or G.M.T: this is an abbreviation for Greenwich Mean Time. Time is calculated from the English observatory at Greenwich. This is located at zero degrees longitude.

Hacking: Rolex introduced this feature in 1973. The feature causes the second hand to stop dead when the crown is pulled out fully to adjust the time. This is perfect if you want to synchronize watches.

Hallmark: A mark or stamp on Gold, Silver and Platinum. All Gold and Silver sold in the UK has a hallmark by law.

Horology: The profession of making timepieces and the science of measuring time.

Jewel/Jewels: A precious stone that is used as a bearing for the watch gears to reduce friction. The top of the jewel normally holds a small amount of lubrication in a depression. Some jewels are set in metal plates that support the movement gears. A pin goes through a hole in the jewel.

Lugs: The two pointed edges on either end of the watch case. This is how the watch strap or bracelet is attached. Sometimes also known in the industry as the watch horns.

Luminova: Organic, non-radioactive, luminous paint now used on the hands and hour markers of a dial. The Breitling Navitimer also has second hands painted in Luminova. Luminova replaced the dangerous radioactive material Tritium.

Lunar phase: A display window with successive aspects of the moon throughout its monthly cycle.

Mainspring: The main spring of a watch or clock that supplies the force of motion to the gears.

Moon Phase: a type of dial showing the changes in the different moon phases. Became popular with boys.

Movement: The working movements of a watch or clock.

Perpetual Calendar: An automatic calendar which takes into the length of a month and can also take into account leap years.

Perpetual Movement: This is another name for the automatic/self-winding watch. It has a rotor that travels 360 degrees.

Quartz Movement: A Quartz watch uses a small battery that replaces the mainspring.

Quick set: Introduced on the Rolex in 1971. This new feature allowed the date to be set rapidly via the crown, without having the hour hand pass over the 12 o'clock setting. *The hands don't have to go around and around to set the date.*

Radium: Radioactive luminous material first used on watch hands and dials in 1914. This was replaced by Tritium in the early 50's.

Rotating Bezel: Used on many Breitling watches and the Rolex sport models such as divers or aviator models. It can be used as a calculator, different time zones and checking decompression times.

Rotor: The weight that is used to self-wind automatic watches.

Sapphire Crystal: Used on most watches today, they are also scratch resistant.

Scale: Graduated marks along the dial, bezel or the exterior of the case, that measure speed and distance.

Serial Number: The identification number on the watch, normally between the lugs, inside or outside the back case. This number can determine the date of manufacture and type of watch.

Sweeping Movement: Refers to the movement of the second hand ticking at approx 7 times per second.

Titanium: An extremely resilient corrosion resistant light colored metal, that is very hard and infact now due to problems producing the material, its more expensive than Platinum.

Triplock: This is the screw down crown, fitted Rolex Sea Dweller, Submariner and the Breitling Super Ocean. It was first patented by Rolex in the early 1950's; The Rolex crown with this has 3 dots under the coronet. The Breitling version is very similar in design, but features a screw thread before the crown is pulled out.

Tritium: Luminous material used to paint the hands and hour hands of watches. This was used from the 1950's on Rolex and Breitling. It was discontinued by Breitling in 1996 and discontinued by Rolex in 1998. They now use a safer material Luminova. Modern watches produced with Tritium will be marked 'T', "T 25'" or "T<25" around the 6 position on the dial.

Twinlock: Rolex screw-down feature crown. This has a twin seal against moisture and dust. It was patented by Rolex in 1953. The coronet on the crown (winder) has a horizontal line under.

Vibration: The oscillating movement of a piece from one extremity to the other. Two vibrations make up an oscillation. A normal balance watch produces 5 vibrations per second, this corresponds to 18,000 per hour or higher.

Spelling: The spelling in this book is for the United States, I apologize to my British readers.

US	UK
Caliber	Calibre
Center	Centre
Color	Colour
Tire	Tyre

Some facts *or fiction* you may not know!

1. *An old Italian Mafia Don is dying and he called his grandson to his bed, " Grandson I wanta you to listen to me. I wanta you to take my 45 automatic pistol, so you will always remember me". "But grandpa I really don't like guns, how about you leaving me your Gold Rolex watch instead?".*

"You lisina to me, some day you goina to be runna da family bussiness, you goina have a beautiful wife, lotsa money, a biga home and maybe a couple of bambino, some day you goina come home and maybe finda your wife in bed with another man. Whata you gonna do then? Pointa to you watch and say, "TIMES UP"?

2. *It was opening night at the Theatre and the Great Lanzero was topping the bill. People came from miles around to see the famed hypnotist do his stuff. As Lanzero took to the stage, he announced, "Unlike most stage hypnotists who invite two or three people up onto the stage to be put into a trance, I intend to hypnotize each and every member of the audience." The excitement was almost electric as Lanzero withdrew a Breitling pocket watch from his pocket. "I want you each to keep your eye on this Breitling watch. It's a very special watch. It's been in my family for four generations." He began to swing the watch gently back and forth while quietly chanting "Watch the watch, watch the watch, watch the watch, watch the watch. You will do as I command, watch the watch..."*

The crowd became mesmerized as the watch swayed back and forth, light gleaming off its polished Gold surface. Hundreds of pairs of eyes followed the swaying watch, until suddenly it slipped from the hypnotist's fingers and fell to the floor, breaking into hundreds of tiny pieces. "Shit!" said the hypnotist.

It took nearly two weeks to clean up the theatre after that.

3. *A girl was visiting her blonde friend who had acquired two new dogs, and asked her what their names were.*
The blonde responded by saying that one was named Rolex and one was named Breitling.
Her friend said, "Whoever heard of someone naming dogs like that?"
"Hellooooo," answered the blonde . "They're watch dogs!"

4. *An attorney opened the door of his BMW, when suddenly a speeding car came along and hit the door, ripping it off completely. When the police arrived at the scene, the attorney was complaining bitterly about the damage to his precious BMW. "Officer, look what they've done to my Beeeemer!" he whined.*

"You attorneys are so materialistic, you make me sick!!!" said the officer, "You're so worried about your stupid BMW, that you didn't even notice that your left arm was ripped off!!!"

"Oh my god", replied the attorney, finally noticing the bloody left shoulder where his arm once was, "Where's my Rolex? "

5. *Tommy has stolen the rabbi's gold Rolex watch. He didn't feel too good about it, so he decided, after a sleepless night to go to the rabbi.*
'Rabbi, I stole a gold watch.' 'But Sammy! That's forbidden! You should return it immediately!'
'What shall I do?' 'Offer to give it back to the owner.' 'Do you want it?' 'No, I said return it to its owner.' 'But he doesn't want it.'

'In that case, you can keep it.'

6. *Three salesmen were bragging who is the best.*

The first said that he is so good he sold a color television to a blind man.

The second bragged he sold a HI-FI stereo system to a deaf man.

The third said he sold a Breitling Cuckoo clock to a blonde lady.

The other two said, so what? And the third salesman added,
Along with the Breitling Cuckoo clock, I also sold her
one hundred pounds of bird seeds!!!!!

Photographic Credits.

Most images taken by the author, some from Public domain.

British Royal Family Official Press release.

Advertisements where taken from the following sources:

UK, Daily Mail + Financial Mail supplement.

UK, The Mirror

UK, The Sunday Times supplement

UK, Financial Times

UK   The Times

National Geographic Magazine

Fortune Magazine

Men's Health Magazine

Cigar Aficionado Magazine

Geographical Magazine

Flex Magazine

Property Trader

The last word:

Hans Eberhard Wilhelm Wilsdorf. *March 22ⁿᵈ 1881-July 6ᵗʰ 1960* To many, simply known as
Mr. Rolex

Circa 1953

Hans Eberhard Wilhelm Wilsdorf's passion in life was his watch company. Did he intend his watch to make the world a better place? Through its uses, its accuracy and its beauty, probably. He wanted his watch to be considered a symbol of achievement, not a symbol of status and became agitated if it was referred to as such. He made it very well known that it was NOT a status symbol. Wilsdorf's watch was priced "just out of reach of the working -class man. So to earn one, you would have to contribute a bit more, work a little harder and reach a bit farther, be a high achiever, a man/woman with ambition.

A Rolex watch is generally significant of a major achievement in its wearer's life. After Hans Wilsdorf's death in 1960, he owned 85% of Rolex and the Company today is still privately held and overseen by the Hans Wilsdorf Foundation which is a Charitable Organization. The proceeds discretely go to Children's Charities around the world (Mr. Wilsdorf was an orphan himself after his parents died when he was just 12).

If you wear a Rolex, Breitling, Ball, Cartier, Longines, Patek Philippe, IWC, Omega or Zenith You should be extremely proud of that watch on your wrist, and your achievements to get it.